THE CELLULAR CONNECTION

THE CELLULAR CONNECTION
A Guide to Cellular Telephones

Fourth Edition

ROBERT A. STEUERNAGEL

A Wiley-Interscience Publication
JOHN WILEY & SONS, INC.
New York / Chichester / Weinheim / Brisbane / Singapore / Toronto

This book is printed on acid-free paper.◎

Copyright © 2000 by John Wiley & Sons, Inc. All rights reserved

Published simultaneously in Canada.

For ordering and customer service, call 1-800-CALL WILEY.

Library of Congress Cataloging-in-Publication Data:
Steuernagel, Robert.
    The cellular connection / Robert Steuernagel. — 4th ed.
      p. cm.
    ISBN 0-471-31652-0 (pbk. : alk. paper)
    1. Cellular telephone systems. 2. Cellular telephones.
  I. Title.
  TK6570.X6S743 2000
  384.5'3--dc21

                                    99-2'916
                                      CIP

10 9 8 7 6 5 4 3 2 1

CONTENTS

PREFACE

The need to keep in touch — we all have it, whether for business or personal reasons. Now you can keep in touch on the way to work or the shopping center, from a construction site or the golf course, or in a rental car or on your boat.

If you spend time away from your best communication tool — the telephone — a cellular phone can add several business hours to your week. Now you can stay in touch with your office, your customers, or your family, even as you inch along in a traffic jam. And cellular's excellent audio performance ensures clear voice reception.

In today's fast-paced world, the average business manager spends fourteen work-weeks per year on the telephone. Salespeople, doctors, and wheeler-dealers report that when they spend time on the road, a cellular phone makes them much more productive. So it's no surprise that cellular has grown to more than 50 million subscribers and is expected to exceed 100 million by 2005.

Cellular phones are not only for the person who spends time in a car. The first cellular phones — "mobile" phones — were bulky and meant to be mounted in a vehicle. The development of convenient, small portable models has increased the number of applications for cellular phones by making them usable anywhere conventional phones are not available. Architects, carpenters, or electricians working at construction or repair sites can be in touch with their home offices. Newspaper reporters and other journalists on assignment can transmit their stories to headquarters immediately or send written copy from a portable computer via their portable phone.

And let's not forget about the family—its safety, pleasure, and convenience can be greatly increased by using a cellular phone, both in the car and away from it. And the greater affordability and availability of cellular service, as well as the cellular phone, puts it within the average person's reach. Driving to the movies, taking a bus to a sporting event, or just walking to the corner store, instant communications can be at most people's fingertips.

Although using cellular service might be as easy as pushing a few buttons, it is still expensive compared to regular telephone service, and there are several things you should understand when choosing your phone and the service you use with it. Everything you need to know is explained here in an easy-to-understand manner.

You will learn how the cellular system works; what the purposes and advantages of each feature are; how to choose a telephone and telephone carrier; how to install and operate your phone; what to do when traveling outside your home area; how to tap into other communications systems; and what's coming in the future. Also included are informative illustrations, photographs, and a glossary of terms for easy reference.

Welcome to the world of cellular telephones—communications on the move.

ROBERT A. STEUERNAGEL

Carlsbad, California
July 1999

THE CELLULAR
CONNECTION

1

AN INTRODUCTION TO CELLULAR PHONE SYSTEMS

The first mobile communications system began in 1921 when the Detroit Police Department installed two-way radios in its cars. The equipment was bulky, awkward, and a drain on automobile batteries, but it proved its worth. Soon police and fire departments throughout the country were installing two-way radio equipment in their fleets.

Eventually someone realized that private individuals could also benefit from being able to talk with the outside world from their cars, and the first commercial mobile telephone service was instituted in St. Louis, Missouri, in 1946. Early mobile phone service was more like using a radio than a phone. You spoke into a microphone or special handset, and the voice from the other end of the connection usually came through a loudspeaker. You couldn't dial a call; instead the mobile operator at the other end of your radio link established the connection for you. It could take some time before you got through, since only a few radio telephone channels were available in any city, and you frequently had to wait for a free one.

When you finally did get through, your conversation was clumsy, at best. Because of the nature of the equipment, you and the party you were speaking with had to take turns talking, pushing a button before you spoke. If you tried to say something while the other party was speaking, they wouldn't be able to hear you and you wouldn't be able to interrupt them. You had to wait until they decided that they had had their say and gave you the go-ahead to speak—a very frustrating experience, as you can well imagine.

Still, it was worth putting up with such hardships to be able to accomplish needed business while on the road or just to enjoy the luxury of chatting with friends as you motored from place to place. Your car, with its impressive buggy-whip antenna, marked you as a very important person. The concept of mobile phones as status symbols is illustrated by the legendary story of the Hollywood executive, who, while speaking to an associate from his car, asked him to hold on for a moment—the executive's *other* phone was ringing.

Improvements in the design of electronic equipment soon made mobile telephones easier to use. In 1948, the first automated mobile dialing system was demonstrated, although it was not used commercially until fifteen years later. For the first time, you didn't need the assistance of the mobile operator—you could place your calls

"Would you please hold for a moment? My other phone is ringing . . ."

yourself. And you could even dispense with push-to-talk operation and converse almost normally. You still got plenty of busy signals, though, waiting for a channel to open up in a congested area.

In 1969 the Improved Mobile Telephone Service (IMTS) was introduced, although it was not much of an improvement over what had existed before. The number of channels was still limited, and in some areas there were waiting lists for mobile phone installations stretching into years. The range of a particular system was also limited to a radius of between 20 and 25 miles from the centrally located transmitter, and interference from other phone systems was a common problem. Even so, the demand for mobile phone service was greater than could be met.

In the late 1960s and the 1970s there was a growing awareness of just how inadequate the existing mobile telephone service was, and a search was instituted for a better way. A proposal was made at the end of 1971 for a type of service called *cellular* (the concept for which had existed at least as far back as 1947), and in 1978 a trial cellular service began in Chicago, serving about 2000 customers. Within a year and a half, AT&T had created a subsidiary called Advanced Mobile Phone Service, Inc. (AMPS) to develop and market cellular telephone service nationwide, and in October 1983 the first such commercial service was inaugurated in Chicago. The AMPS acronym survives today as the name of a standard under which cellular is designed for use in North America, so that all local systems and telephones work together correctly from city to city.

In 1983 cellular telephones were designed primarily for in-car use, and cost over $3000. The car phone rapidly became less expensive, and advances in technology soon allowed the portable phone to become practical and affordable. By 1991, most cellular systems had been installed across the country in rural and suburban areas as well as cities, and the portable had become more popular than the vehicle-installed version.

THE CELLULAR METHOD

Until the advent of cellular phones, radiotelephone systems—even IMTS mentioned earlier—worked pretty much the same way. In your car you had a radio transmitter and receiver (the combination is called a *transceiver*) and at a point central to your service area was

A popular portable cellular telephone in the "flip-phone" style. (Photo courtesy of Motorola, Inc.)

another, more powerful transceiver, operated by the telephone company to which you subscribed. This area transceiver could connect you into the regular telephone lines, through which your conversations with the rest of the nonmobile world took place.

An area or city was served by a single transceiver location. The IMTS transmitter had a range of 20 to 25 miles using a power of perhaps 250 watts. As you got farther away from the central antenna location, signals—both to and from the central site—became weaker and noisier, often making it difficult to maintain a conversation. In addition, if you were in a region with a number of mobile phone services, there was a good chance your conversation would be interrupted by interference from other mobile users. You were generally restricted to using just the service you subscribed to, and, if you ventured outside your local area, your expensive mobile phone installation became useless.

But cellular service operates in a completely different fashion. Instead of having one central, high-powered transmitter to cover an entire region, the cellular system divides that region into a number of small "cells" just a few miles across, like the cells in a honeycomb. Although these cells are circular because of the nature of radio signals, which radiate in all directions from a single source, they are usually represented on maps and in drawings as hexagons, since that makes it easier to show graphically where one cell ends and another begins (Figure 1.1).

Each cell has at its center a cell site, where the fixed radio

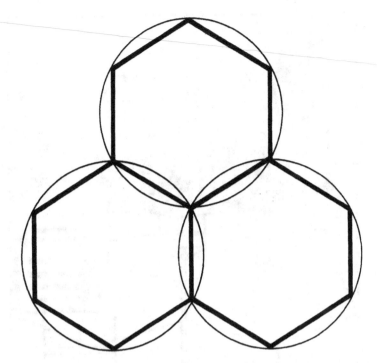

Figure 1.1   CELLS ARE REPRESENTED BY HEXAGONS
Cellular service areas, or cells, are actually circular in shape because of the nature of radio waves. However, they are usually represented by hexagons, since that makes it easier to show where one cell ends and another begins.

transceiver is located. All the cell sites belonging to a particular system are connected together at a Mobile Switching Center (MSC), also called a Mobile Telephone Switching Office (MTSO), which ties them into the conventional phone system. The transmitter at the cell site is low power (100 watts or less), and the effective useful radius of a cell is only a few miles. When you approach the working limit of one cell, your call is transferred, or "handed off," to a cell site closer to you that can "hear" you better (Figure 1.2).

The cellular telephone you use also has much lower power than the ones used in older, single-transmitter systems—3 watts maximum, or 0.6 watt (600 milliwatts) for portables. This limited-range cellular approach offers quite a few advantages over those used for earlier mobile systems. A number of small cells means that when you are within an area of cellular service, you are always assured of strong

Figure 1.2   HOW CELLULAR PHONE SERVICE WORKS

Each cell has at its center a cell site where the fixed radio transceiver is located. All the cell sites belonging to a particular system are connected together at a mobile telephone switching office, which ties them to the local phone system. As you pass from one cell site to another, your call is transferred or "handed off" to the next cell without any noticeable interruptions.

signals — as you approach the limit of the range of one cell, you will enter another before the quality of service deteriorates.

Furthermore, the limited range of cellular transmissions means that there will be no interference from other neighboring systems and that the same frequency can be used again just a few cells away. This is called frequency *reuse*, and makes it possible for the cellular system to provide greater capacity within a single metropolitan area. Finally, the cellular approach allows for easy expansion of service. When a particular area becomes saturated to capacity, a cell can be "split" to form several new cells that provide more capacity in the same area, each with its own cell site. This can also be done by adding directional antennas to a cell site so it can radiate several noninterfering signals in several directions at once, on frequencies different from the ones used by the original site. Each newly created cell can handle up to about sixty channels. Newer techniques include "microcells," which provide even smaller cells to reach hard-to-cover areas and still more reuse and capacity.

In the chapters that follow, we'll examine in greater detail how cellular systems — and, in particular, cellular phones — operate, and present you with the information you'll need to take advantage of the many pluses of cellular communications.

Now you can stay in touch with office, clients, and family while on the road. (Photo courtesy of Motorola, Inc.)

2

CELLULAR PHONE EQUIPMENT

There are three parts to a cellular phone system: the cellular telephone, or cellphone; the cell site, which receives and transmits radio signals from and to your phone; and the MSC, or Mobile Switching Center, which links cellphones to established conventional telephone services. The relationship among these is illustrated in Figure 2.1.

THE CELLULAR TELEPHONE

The standard portable cellular telephone is a one-piece unit that serves as the handset, and includes a dial pad or "keypad," display, microphone, miniature speaker or earpiece, a miniature radio receiver and transmitter or transceiver, antenna, and a removable, rechargeable battery. The small portables available are marvels of engineering, and may have a standard, rectangular shape, or the popular flip-phone styling, which covers the keypad when not in use and is reminiscent of the "communicator" used in the *Star Trek* television series and films. Additional, unique styles provide different hand-fitting shapes and additional electronic features. The Motorola Classic series, for example, is noted for its durability under rough

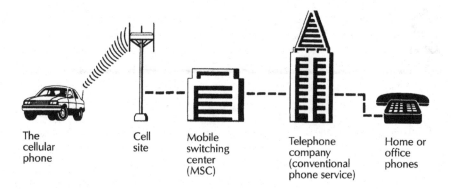

Figure 2.1   PARTS OF A CELLULAR SYSTEM
*The three parts of a cellular phone system are [1] the cellular phone; [2]
the cell site, which receives and transmits your calls to and from your
phone; and [3] the mobile switching center, which then transmits your
calls to conventional phone service.*

conditions and long battery life, rather than small size, and is
nicknamed "the brick" because of its unique size and shape.

These portable telephones have reduced power output of 0.6 watt,
or 600 milliwatts, compared to 3 watts in standard mobile cellular
phones, in order to reduce the battery size and thus the overall size
of the unit. This power reduction may only produce noticeable
reduced performance in fringe areas of cells, in buildings, or inside
vehicles, and is not a serious problem for the average user. Power
boosters and installation kits are available for use in vehicles, which
not only increase the power of portables in the vehicle, but provide
a convenient cradle in the vehicle and provide "hands-free" operation
with a remote speaker and microphone installation.

Portable phones, of course, carry with them their own power
sources, usually a rechargeable battery pack. Some have special
battery holders for regular alkaline batteries, which are held in
reserve in case the phone is needed when the rechargeable battery is
discharged. The battery may be the standard nickel–cadmium var-
iety, or the higher-capacity, more compact NiMH (nickel–metal
hydride) or lithium-ion type. The phones are supplied with a charging cradle, which often will charge the batteries alone or installed in
the phone, as a convenient place to keep it when not in use. Several
types of chargers are available, which may provide rapid charging,
small size for travel, and other advantages.

Portable phone with self-contained battery and antenna. (Photo courtesy of Nokia.)

Portables have self-contained antennas, usually flexible, "rubber duckie" types that are either fixed, or extendable where necessary for better reception.

We will discuss other types of phones, including larger portables, called *transportables*, and car-installed phones, in a later chapter.

When most cellular systems were new and cell sites were far apart, portable service was noticeably spotty because of the portable's low power. In-building and in-auto use were particular problems. Today, most cellular systems are built out to the point where no difference in performance can be detected between portable and full-power mobiles in most situations.

Thus, more than 90% of new buyers are opting for portables; they are more convenient, can be used in offices and other areas out of the car, and are easily carried along anywhere for business trips. However, they are easy to misplace, and are a major source of irritation to their owners because most only allow about an hour or two of "talk time" between battery charges, although this is improving all

Another style of portable phone. (Photo courtesy of Audiovox Corporation.)

the time. A second battery is a must for most users, for use while the other is charging.

WHAT YOU SHOULD KNOW ABOUT CELLULAR FREQUENCIES

The frequencies used by cellular telephones range from 824 to 894 megahertz (MHz), with a gap between 849 and 869 MHz that's used by other communications services. (A map of the 800-MHz cellular

A typical cellular telephone installed in a car. (Photo courtesy of Nokia.)

phone spectrum is shown in Figure 2.2.) Some of these frequencies were originally assigned to the top portion of the UHF-TV spectrum and were intended to be used by TV translator services for relaying commercial TV signals to small rural communities that had difficulty receiving regular broadcast services. The cellular phone frequencies are divided into two bands, and each band is subdivided into two sets of adjacent blocks, A and B.

Each area of cellular service was intended to be serviced by two companies—a *wireline* service (a telephone company affiliate that usually already handles the existing landline or wire telephone service in the area) and a *nonwireline* service (one that is usually already involved in other types of mobile radio communications or that operates a paging service). The differences between these two types of

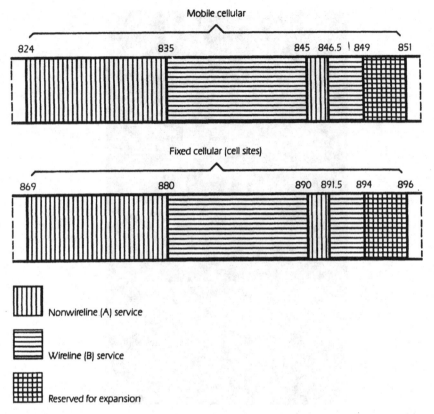

Figure 2.2   CELLULAR FREQUENCIES
*Cellular frequencies range from 824 to 894 MHz, with a gap between 849
and 869 MHz for use by other communication services. The frequencies
are divided into two bands, and each band is divided into two sets of
adjacent blocks, A and B. Half of each band is assigned to a nonwireline
[A] service, and the other to a wireline [B] service.*

carriers are discussed in greater depth in Chapter 3, "The Business of
Cellular Phones."

One block of each band, the A block, was assigned to a nonwireline, and the other to a wireline, the B block. The lower band
(824–849 MHz) is for use by cellular phones to transmit to cell sites,
and the upper one (869–894 MHz) is for use by cell sites to transmit
to cellular phones. There are 832 frequencies, or channels, allocated
by the Federal Communications Commission (FCC) for cellular use;
each carrier — wireline or nonwireline — is assigned the use of half of
them. Each transmit or receive channel is 30-kilohertz (kHz, one

MHz equals 1000 kHz) wide, which provides plenty of room for high-quality audio transmission and reception, with guard bands between channels to prevent interference. The new Personal Communications Service (PCS) services operate at much higher frequencies, that is, the 1900-MHz, or 1.9-*gigahertz* (GHz) band.

NUMERIC ASSIGNMENT MODULE (NAM)

One of the most important and most interesting parts of a cellular phone is a small integrated circuit, or chip, called a NAM, for Numeric Assignment Module (NAM rhymes with "Pam"). The NAM chip is programmed, usually by your cellular phone dealer or installer, to contain the information that uniquely identifies your phone to a cellular system when you place a call or when someone is trying to reach you; this information includes your new cellular telephone number.

The process of NAM programming, sometimes call *burning* the NAM, requires special equipment or key sequences on the phone, and once information has been entered into a NAM it usually cannot be changed without this special equipment. Some NAM programmers are specialized "dedicated" devices, and some are intended to be used with a personal computer. Most new cellular phones can program their own NAMs using special key sequences.

Included in the information programmed into a NAM is the serial number of the cellular telephone you are using and the phone number assigned to it, along with other information. The computers at cell sites and MSCs use this information to identify you when you use the phone — which, among other things, helps the cellular company in preparing its bills, identifying you as an authorized user — and to locate you when someone calls you.

The information contained in a NAM personalizes the equipment that contains it, which also makes it useless to someone who steals it, and useful in identifying the owners of stolen cellular phones that have been recovered.

A QUICK VISIT TO A CELL SITE

A cell site, which may cost half a million dollars or more to construct and equip, is the link between your cellular phone and the rest of the cellular telecommunications system. It is where the messages bound for you leave the ground, as it were, and is the first stop for calls coming from your phone.

To get the best coverage, a cell site is sometimes located atop a tall building in metropolitan areas or on a high point or mountain in less built-up areas (Figure 2.3). This allows it to have the best radio "view" of the territory it is responsible for. Just as often, a cell site will be purposely located in an area of low elevation in a densely populated city area, in order to provide a lot of capacity, or channels, to a very small area, with minimum interference to adjacent cells. In

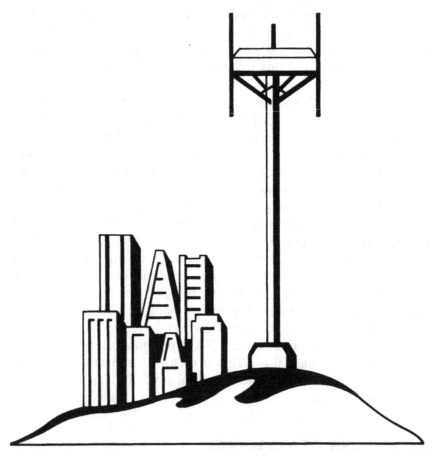

Figure 2.3   CELL SITES NEED A GOOD RADIO "VIEW"
Cell sites are sometimes located atop a mountain or tall building to gain the best radio "view" of a territory.

the high-frequency band that cellular occupies, radio waves propagate only along a line of sight (and as reflected from obstructions), and this makes the siting of the antenna very critical. Unlike your cellular telephone, the cell site usually uses several antennas, each beaming and listening for signals in a different direction. These

You can recognize a standard cell-site antenna tower by its characteristic triangular antenna array. (Photo courtesy of Valmont/Microflect.)

directional antennas ensure optimum results within a cell's area of coverage and may be individually adjusted for best results.

Each antenna actually consists of two antennas—one for transmitting and one for receiving. A single such unit is capable of handling any number of two-way conversations on different frequencies simultaneously.

Inside the cell site are the transmitters and receivers connected to the antennas, along with the equipment necessary to monitor the operation of the cell site and keep it in proper working order. Also located at the cell site are the electronics that connect the cell site to the MSC (usually by coaxial cable, microwave, or fiber-optic link).

A cell site can handle up to about sixty channels, based on using the same frequency used by another cell site that is no closer than seven cells away ($7 \times 60 = 420$, or approximately the 416 channels allowed one carrier). Reuse of these same channels by nonadjacent cell sites within the same area, provides greater communications-handling capability while reducing a cause for interference between neighboring sites. For example, if you are using channels A and B at cell site one, the mobile phone users in the cell site next to you might use channels C and D. But, the people *seven* cell sites over will be able to "reuse" channels A and B, since they will be far enough away from you to do so without interference.

As cellular service continues to grow, more and more cell sites appear at closer intervals, often down to less than one mile apart in densely populated city areas. Where cell sites are nearer to each other, there is less need for height. Rather, the antenna is kept low to reduce interference with other cells.

While height reduction, careful zoning, and careful design are used to prevent the antennas from becoming eyesores, they are becoming numerous. Standard cell-site antennas are easy to spot because of their characteristic triangular array.

The latest developments in cell-site antennas permit them to reuse frequencies in even closer proximity—sometimes every third cell site. New cell sites called *microcells* reduce the radius of cells in congested areas to less than one-quarter mile. Also, structures that disguise the antenna as a tree or church steeple, while providing full transmitting capability, are now feasible. Using electronics, "smart" antennas can dynamically change the pattern the radio waves propagate from the antenna. Both the cell-site radio and the cellular phone are capable of dynamically adjusting their power level. It can be raised to improve the signal, or lowered to reduce interference.

3

THE BUSINESS OF CELLULAR PHONES

The business of cellular phones is a big one. Although cellular phone companies must make a large initial investment, the potential for an enormous return makes it certain that there will be no shortage of firms to service the cellular user.

WHERE CAN YOU GET CELLULAR SERVICE?

Cellular service was deployed beginning in 1983 in order of city population, with licenses for the largest cities awarded first.

While some potential users have had to wait for service in smaller cities, in 1992 the last tertiary markets established commercial service. Now, cellular service is available in all populated areas in the United States (including Alaska and Hawaii), Canada, Mexico, and Puerto Rico. Only the most desolate wilderness areas may not have complete coverage.

HOW LICENSES WERE GRANTED: WIRELINE AND NONWIRELINE COMPANIES

The Federal Communications Commission (FCC), which controls and watches over the cellular communications industry, is responsible for issuing the licenses required to operate cellular systems.

To encourage competition, the FCC decreed that provision be made for each area to be serviced by two cellular phone carriers, one a wireline and the other a nonwireline service. The *wireline* carrier is usually affiliated with the organization that already provides conventional telephone service in the area — your local telephone company. Such companies are usually already involved in mobile communications in areas other than cellular and frequently bear a name similar to that used by their parent organizations. The wireline carrier in an area was assigned the B block of frequencies (see Figure 2.2).

The competitive component required by the FCC in the cellular market was provided by another sort of company, a *nonwireline* service. This company was usually not directly involved with supplying standard phone service, but was usually involved in providing two-way mobile communications, paging, or another radio-based service in its region of service. The nonwireline carrier was assigned the A block of frequencies.

In the case of a wireline carrier, there was usually no competition for the license in a particular area, and the monopoly phone company was awarded the license. In the case of the nonwireline license, there were several competing companies who applied for the license, and there was some delay until the license was awarded. Wireline carriers were thus usually first in providing commercial cellular service in their area. In smaller markets, there was so much competition for the nonwireline license that the FCC chose to use a lottery system to award licenses to nonwireline applicants.

Because wireline and nonwireline companies have been buying and trading each other's properties since the licenses were awarded, these distinctions originally established for the purpose of awarding licenses have been blurred or have disappeared altogether. Conventional telephone company affiliates own the A block license in some cities, and the B block license in others. The nonwireline company, McCaw Cellular, sold its large complement of licenses to AT&T in 1994, putting this telecommunications giant back in the cellular service business after many years of being out of it, as the largest carrier of *nonwireline* cellular service in the country.

Some of the largest and most well-respected companies in the world are involved in some aspect of cellular. Wireline service providers include regional Bell telephone company Bell Atlantic, which has merged its cellular operations into Bell Atlantic Mobile after buying NYNEX; BellSouth, AirTouch (separated from Pacific

Cellular phones also have many special applications. This emergency cellular call box is powered by solar cells and permits emergency calls. It requires no telephone or power lines to be installed. (Photo courtesy of Connectivity, Inc.)

Telesis and merged with US West's cellular operations), Southwestern Bell, now known as SBC, and GTE; and long-distance carriers AT&T, Sprint, and MCI (as a reseller only). The equipment used by carriers to build cellular systems is provided by AT&T Network Systems Division (now known as Lucent Technologies), Motorola, Ericsson, NorTel, and Nokia, among others. Cellular telephones are manufactured by some of these companies as well as Japanese companies like Panasonic (Matsushita), Mitsubishi, and Sony.

WHO SELLS CELLULAR PHONES

Cellular telephone equipment is available through many types of dealers—from the local cellular carrier's own stores, or through its authorized agents, as well as independent dealers and retailers. Traditional cellular dealers usually have a garage to do vehicle installations, and a major technical service facility. Since most phones sold are now portable and do not require installation, a special kind of facility is not required. Authorized agents and dealers include not only consumer electronics stores and car stereo installers, but more nontraditional outlets like kiosks in malls, the electronics counter at a drugstore, or even by ordering by phone, by mail, or the Internet.

In addition to selling you your phone, programming it, and arranging for service to be started, most dealers will either install a mobile phone or car hands-free kit for your portable, or will refer you to someone who specializes in such installations.

The best way to find out about cellular facilities in your area is to ask the people who own and use the phones. They'll be proud to show them off to you and will be a rich source of information on phones, service, and the cellular situation in general.

Much more on the subject of selecting and purchasing a cellular phone can be found in Chapter 4, "Getting Cellular Service."

4

GETTING CELLULAR SERVICE

Cellular phone equipment must meet stringent standards and pass a number of stiff tests, so you are assured of getting a quality product no matter where you buy it. The real deciding factor on where to buy your cellular phone is the human one—the standards, know-how, and customer service capabilities of the dealer.

FINDING THE BEST SOURCE OF CELLULAR PHONES AND SERVICE

Cellular phones are a hot item. In addition to Yellow Pages advertising, you can look in any newspaper; ads for cellular equipment and services abound. Advertising for cellular service is also found on radio and television, and increasingly in national as well as local commercials, because of the consolidation of carriers and the participation of national companies like AT&T.

Choosing the right cellular carrier is just as important as choosing the right cellular phone. Across the country, more and more dealers are discounting the price of a cellular phone to almost zero for the most basic models, and less than wholesale for the more sophisticated ones. Dealers can afford to do this because they use the sales commissions paid to them by the carrier for cellular subscriptions to "buy down" the retail price of the cellular phone, so that the price of

Table 4.1   The Cellular Buyer's Checklist

1. The phone I've selected is just right for my needs, easy to use, and has all the features I need at a reasonable price . . . .
2. The carrier I've selected has good customer service, coverage, and the service features I need . . . .
3. The rate plan I've selected is right for the place, time, and amount of calls I expect to have . . . .
4. I understand how I'll be billed and the items I'll see on my first bill . . .
5. I understand the terms of my service contract and warranty . . . .
6. My phone is working and I know how to use it . . . .

the phone is not an obstacle to your subscribing to cellular. However, the dealer doesn't receive a commission unless you subscribe with a one- or two-year service contract for cellular service with the carrier that he or she represents. This contract assures the carrier that you will be a customer long enough to justify the commission or low telephone price you are paying. There are penalties for ending the contract early.

Since virtually all dealers work this way, almost any cellular phone is a bargain, as long as it has the features you need and you subscribe to the service when you buy your phone. Selecting the right rate plan from the right cellular provider is the most important thing for you to do to save money and get the most value from cellular service. Rate plans, also called pricing or service plans, are discussed in detail in Chapter 5, under Illustrative Pricing Plans.

No matter where you go to buy cellular, it can be complicated. In order to make it as simple as possible, we've included a checklist for cellular buyers (see Table 4.1). You should probably take it with you when you go.

EQUIPMENT COSTS

The price of cellular phones has fallen dramatically since they.were introduced. Phones that used to cost $3000 are now available at subsidized prices under $100, or virtually free. Premium models that have many features and digital service can still cost several hundred dollars, as can one that includes a power booster and hands-free kit for the car.

All cellular phones are heavily discounted, and only the most sophisticated models will cost more than $200, whether portable, mobile, or transportable. Installation of a car phone, or a hands-free kit for using a portable in a car, may cost $100–$300. Additional accessories, which we will cover in more detail later, include such items as carrying cases and extra batteries, and may actually cost more than the phone itself.

Since most phones are a bargain anyway, it is worthwhile to invest in a premium phone with the right accessories to make it more valuable as a tool and more convenient to use.

In the past, when they were very expensive, many people chose to lease or rent their phones. With the lower prices of phones today, these options are no longer necessary. Rental phones are still popular as temporary phones for rental cars, conventions, and other special

events, or when out of town. You may want to consider this option to try a cellular phone without incurring a service contract obligation, but remember that premium usage charges usually apply. Major automobile rental agencies often rent out cellular phones, and are a good first place to look for a rental phone.

WHAT TO LOOK FOR IN A CELLULAR PHONE

Cellular telephones, because of their microprocessor-based design, include a number of features often not included except on the most expensive conventional telephones, and are of an extremely small size. These features include a display and telephone number memory, for example, as well as many others designed exclusively for cellular and portable operation, including a battery-level indicator and signal-strength indicator. These hardware and performance features are discussed in more detail in later chapters. For the moment, let's examine a checklist of things to bear in mind when shopping for a phone (Table 4.1).

First, make a complete list of all the expenses that will be involved in operating your phone. These include the cost of the phone itself, the accessories (including such basics as an extra battery and charger), any installation required, and the subscription and use of a cellular service. The next chapter presents some information about operating costs.

If you intend to use the phone in a cellular service fringe area, consider a car-installed phone or a power booster kit designed for the portable phone you choose. Determine the repair policy of the dealer. Many will provide you with a loaner phone if your phone needs repair, or even replace it rather than repair it.

The main concern for choosing a cellular phone is how you will use it. A phone used mainly for emergencies and light personal use requires only the most basic features; premium features may even make the phone more complicated to use for the infrequent user in an emergency.

Regular users of portable phones will require a phone that has a reasonable amount of "talk time" (using the phone for calls and actively transmitting and receiving) and standby time (turned on, but just waiting for incoming calls). This is measured in minutes of talk time and hours of standby time. Regular users should pay a premium

A modern wireless store. (© 1999 PrimeCo Personal Communications, L.P.)

for a model that has talk time of more than 90 minutes or standby time of more than 12 hours, or for a model that has optional high-capacity batteries to provide this capacity. Such premium batteries may change the size or "feel" of a phone, and this should also be considered.

Heavy business users should consider getting a car-installed phone plus a portable, or a portable with a vehicle installation option and power booster, to make sure service is conveniently available in any situation.

Because phone prices are often subsidized, almost any phone is a great value. Therefore, the placement of the keys, readability of the display, and other features are the most important considerations.

Digital cellular service is available in many areas. Cellular phones capable of digital service are usually also able to use regular, or analog, service, and are called *dual-mode* phones. Digital service provides among other advantages, more secure conversations and longer battery life of the phone. These phones cost more, but provide these advantages, and with the latest technology are less prone to obsolescence, although with today's phone prices so low, this is not a major concern.

Try out the phone you're interested in to see that it will serve you well in your area and that it has all the features you want. (Photo courtesy of Motorola, Inc.)

TRYING IT OUT

Before you buy a cellular phone, you will probably want to try one out. Most dealers will allow you to try a portable phone right in the retail location, or allow you to return it if it does not perform. It is important for new users to adjust to the method of initiating calls and getting used to its small size and buttons. Make sure the phone fits conveniently in the pocket or purse where you will usually carry it, with the size battery you will usually use.

CELLULAR SERVICE PRICING — YOUR MOST IMPORTANT DECISION

After you've chosen the type of phone you need and estimated the cost it will entail, you need to think about how you will be using your

Table 4.2 Estimating Cellular Usage

1. I plan to use my phone for: (1) Mostly business; (2) Mostly personal; (3) Business and personal; (4) For emergencies only . . . .

2. On a typical weekday, the number of calls I'll make are: (1) None; (2) 1–3; (3) 4–10; (4) More than 10 . . . .

3. My calls are usually: (1) Average length, 2–3 minutes; (2) Long, over 5 minutes; (3) Short, less than 1 minute . . . .

4. The proportion of calls I'll make after 7 P.M. and on weekends ("off peak") is: (1) 0–20%; (2) 20–40%; (3) 40–60%; (4) More than 60% . . . .

5. 80% or more of my calls will be made at this distance from my home or office: (1) Within 5 miles; (2) 5–20 miles; (3) County- and state-wide . . .

6. I will be regularly roaming (using my phone in service areas outside my home metropolitan area) . . . .

7. I probably need Call Forwarding and voice messaging to answer calls when I'm not available . . . .

8. I need paging to be available immediately/to avoid incoming calls . . . . .

9. I'll be using fax or data on the go . . . .

Using this table: If you plan to use the service for emergencies only, there is probably a minimum-use or safety plan you can pick right away. If your usage is all within 5 miles of your home or office, there may be a "zone"-type plan that saves you the most money. Some zone plans may restrict roaming, so check the roaming question also if you are a candidate for this plan. As a quick approximation, most users can multiply their weekday calls by the average call length by 20 weekdays a month, then multiply the percentage of off-peak by the result and add it, for example:

$$3 \text{ calls} \times 2.5 \text{ minutes} \times 20 \text{ days} = 150 \text{ minutes per month} - \text{peak}$$

$$150 \text{ minutes} \times 30\% \text{ off-peak} = \underline{ 45 \text{ minutes per month}} - \text{off-peak}$$

$$195 \text{ minutes per month total}$$

phone, where you will use it, and how much you will use it. This will help you estimate how much cellular service will cost each month.

Table 4.2 will help you organize your expected usage characteristics, so your dealer can assist you in choosing features and rate plans. As we mentioned, the dealer where you buy your cellular phone may only represent one service provider. Cellular service pricing is competitive because there are at least two service providers, and more where resellers are active. Each service provider will have rate plans that are similar, but may differ enough to make a big difference in the total price you pay for service.

While each local provider will have unique plans and names for their plans, they are usually variants of four or five basic structures that we discuss in the next chapter, after introducing billing elements and terms.

As we continue to get further into the discussion of the cost of service, you may become concerned that cellular is expensive. But cellular is a valuable service, and it only seems expensive when you compare it to the price of ordinary fixed-telephone service. While cellular has great value for all its users in terms of convenience, peace of mind, and immediacy, business users can usually quantify its usefulness in productivity gains.

While having the phone available if only to make one timely decision per month will justify the cost to many, most business users simply count the hours per month spent productively on the road or on the go. At $40 per hour for your time, for example, saving 20 minutes per day for 21 business days per month will save you 7 hours, or $280, per month, probably much more than your cellular bill. The higher your cellular bill, the more time you're probably using productively. Ask any business user, and they will tell you that cellular is indispensable.

Service providers may have special offers on phones as well as service, and the cost of cellular is always coming down (see "Special Promotions and Pricing Options" in Chapter 5).

REMINDER: THE CELLULAR BUYER'S CHECKLIST

The most important thing you can do when buying a cellular phone and service is to take Table 4.1 (page 24) with you when you go, and make sure you've completed every step. Not only will you get what you need at a reasonable price, the phone and service will be much more valuable to you if you've taken the time to go over everything on the list.

5

THE BILL, PLEASE

The costs of a cellular phone, of other necessities such as a battery charger, and of installation, are one-time expenses. Once you've paid for these items and services, you're done with them.

Fixed costs, though, account for only a part of what it costs to operate a cellular phone. In this chapter we'll look at the other, more important expenses—what it costs you to *use* your phone.

ONE-TIME (NONRECURRING) EXPENSES

Like the price of purchasing equipment, some operating expenses are paid once and never again (unless, for example, you have your phone number taken out of service and then have it restored, or if you move to a new cellular service area). These one-time expenses are outlined in the following text.

Service Activation Charge

This fee, typically about $50, is the service provider's charge for making the electronic arrangements for service to your number and for processing the paperwork associated with it. If you suspend service and then request that it be resumed, you may have to pay this fee again. This fee is sometimes waived during special promotions.

Change Charge

If you ask that your service be switched from one phone to another using the same number (maybe you just bought a new model), you will have to pay for the expenses incurred in making the switchover. Change charges may be around $15.

Deposit

You will be subject to a credit check when you initiate service, for which you will have to grant permission. If your credit is less than perfect, you may be asked to pay a deposit to make sure that you can pay your bill. If you choose prepaid service, the deposit and credit check may not be necessary.

RECURRING EXPENSES

Just as with your home phone, the use of a cellular phone brings with it certain fixed charges, usually the same amount in every billing period. There are also variable recurring expenses, discussed further on. The following are among the fixed charges that repeat from month to month, some of which are optional features.

Access Charge

This is the monthly fee for basic service, and, as explained below, may or may not include a number of free minutes of use, or calling time, called *airtime*. The access charge will vary depending on how much free airtime is included. It may also vary depending on how long a service contract you sign.

Minimum Usage Fee

This is usually part of the access charge. Various rate plans will have a different monthly access charge, depending on how many minutes of airtime use are included with it. You can calculate how much each minute of included airtime is costing in the rate plan by subtracting the access fee for a similar plan that includes no usage, and dividing it by the number of minutes of free airtime. This usage is usually discounted because you are committing to it as a minimum. You will be billed for this minimum amount of airtime whether you use it or not. Special services (see the following subsection) may also be included in the access charge in some plans, in addition to the minimum usage fee. These will also be discussed in Chapter 6.

Features or Special Services

If you have opted to add to the basic cellular service some of the options offered by your service provider, you'll be billed for them on a regular basis. Interestingly, the cost of these additional conveniences (about $3 apiece) is frequently less than that of the equivalent for conventional telephone service. These extras can usually be added to your regular service at any time, and may be available in combinations at a discount. Some of these features may be included free with your monthly access if you choose a premium-rate plan.

The following are among the additional services available.

Call Forwarding   This service permits you to have calls transferred automatically from your cellular phone to your home or office phone, or any other number. If you have Call Forwarding, you can switch it on and off from your cellular phone simply by pressing a couple of keys. A similar feature, called *Conditional Call Forwarding* or *Busy/ No-Answer Transfer*, allows calls to your cellular number to be transferred to the alternate number only if you do not answer, or the phone is busy, or both. This feature is also used in combination with voice messaging to answer your phone when you can't.

Call Waiting   This service gives you, in effect, a second line to your cellular phone. If a call is placed to your cellular number while you are already using your phone, you will be alerted by a beep. You can put the first party on hold to speak to the second, or vice versa. This feature can also be turned on and off with a couple of key presses. It is very useful if you think you will have calls that you don't want to miss. But for many users, the chances are low that you'll ever have an incoming call while you are on the phone, and the feature may interfere with data calls, if you use the phone for on-line computer use. You can use voice messaging (described later) to answer the phone and take a message if your phone is busy.

Local Calls Only This feature prevents your phone from being used for long-distance calls without a calling card — calls can only be placed within your calling area. This can be a useful feature if people other than yourself use your phone.

Incoming or Outgoing Call Restriction This permits you to make your cellular phone *receive only* (people can call you, but you can't call out) or *send only* (you can call out, but no one can call you). These features are usually only for special applications, such as dispatch services for commercial vehicles (Incoming Only) or for use by customers in a taxi (Outgoing Only).

Additional Features Other features, such as voice messaging, which may also be used for telephone answering, local and nation-wide paging, and other services, are also available from some service providers and can be included on your cellular service bill. These features and those just cited are also discussed in other chapters under "Options and Accessories" and "Special Cellular Features" further on in this chapter.

VARIABLE CHARGES AND PRICING PLANS

As you might expect, the more you use your cellular phone, the greater your monthly bill will be. Airtime, or usage, is the major variable recurring expense, as it is charged according to the amount you use your cellular phone, and combined with the monthly access charge already defined, makes up the bulk of the bill. While the access charge is considered a fixed charge, it may be different depending on which pricing plan you pick. Because access and usage are related in most pricing plans, we will combine our discussion of variable airtime charges with the discussion of pricing plans.

The way you are charged for service will vary from one service area and service provider to another. The charges and billing methods described here are intended to be typical of what you can expect; they do not represent the rates or policies of any specific carrier.

Companies offering cellular service have several pricing plans that you can select from. Most of these require that you sign a service agreement, which commits you to a term of service of one or two years and provides a discount or incentive for a longer term. Penalties are assessed if you cancel the contract before its term. The very low prices on cellular phones offered today by dealers are contingent on the signing of this agreement. You may be able to obtain service without a contract, but you will have to pay a much higher price for the phone.

Some of these pricing plans include a number of minutes of calling time in the access charge; other plans separate the two. Some plans carry different rates for usage at the same times of day — one plan may be designed for people who make the greatest use of the phone during business hours, while another may be aimed at nonbusiness users who place the majority of their calls in the evening or on weekends.

TERMS ASSOCIATED WITH RECURRING CHARGES

A few basic parameters determine the charges for airtime usage. *Peak* airtime is generally weekdays from 7:00 A.M. to 7:00 P.M., and is more expensive than *off-peak* airtime, which is all other times including evenings, weekends, and holidays. In addition to airtime, there may be a separate *interconnection charge* of a few cents per minute to connect to the telephone network, or this may be included in airtime charges. *Rounding* is the method used to calculate usage. The length of a call may be rounded up to the next whole minute, as with traditional telephone service, or rounded to the nearest tenth of a minute or second. While seemingly trivial, rounding to the nearest whole minute can increase total airtime charges by 10% or more than when the call is rounded to the nearest second.

The access charge we have already mentioned may start at $15 to $30 per month, and may go as high as $50, with no free airtime included (see Figures 5.1 and 5.2). Generally, the rate per minute of airtime goes down as the access charge goes up — with things tending to balance out in the end. In effect, you pay a higher access charge for more included minutes of talk time and the privilege of obtaining a discount on airtime. This increases your fixed charges per month, in return for a lower total bill if you use the service frequently. When you're shopping for a service, explain how you intend to use the phone, and the sales representative will suggest a plan appropriate for you.

```
                       WESTERN MOBILE SERVICES
                            SAN FRANCISCO
                             BILL SUMMARY

  BILLING DATE 10/10                        BILLING ACCOUNT #14002487

  PREVIOUS BALANCE              46.63
  PAYMENTS                      46.63     THANK YOU FOR YOUR PAYMENT
     BALANCE                                0.00
     MONTHLY ACCESS                        25.00
     OPTIONAL FEATURES                      0.00
     AIRTIME USAGE                         65.00
     LONG DISTANCE                          0.00
     PACIFIC BELL                             00

     ROAMER TOLL                            3.77
     OTHER CHARGES AND CREDITS              0.00
     LATE PAYMENT                           0.00
     FEDERAL EXCISE TAX                     2.80
     STATE 911 TAX                          0.83
     LOCAL TAXES                            0.00
     STATE REGULATORY FEE                   0.27
     BILLING SURCHARGE                      0.00

        TOTAL AMOUNT DUE               |    97.67    |
```

Figure 5.1   A TYPICAL CELLULAR PHONE BILL

As an example, a service may have a pricing plan with a $45 access charge, with calls during peak time costing 45 cents per minute and off-peak calls costing 25 cents per minute. Another plan offers service with a $25 base rate or access charge, with calls during peak time costing 90 cents per minute and off-peak calls costing 20 cents per minute. Depending on your pattern of use, one plan or the other will be more to your benefit. Business users would generally find the first plan less expensive, while nonbusiness users would favor the latter.

ILLUSTRATIVE PRICING PLANS

There are usually three to five types of rate plans or pricing plans, each of which may be attractive based on usage or contract length. We discuss the general types here.

- The *basic* plan is really a benchmark or yardstick for measuring the benefits of all other plans, and has no contract attached to

Figure 5.2   Message Detail from a Typical Cellular Phone Bill

WESTERN MOBILE SERVICES
SAN FRANCISCO

MESSAGE DETAILS FOR (415) 555-0000

						AIRTIME	LANDLINE
CHARGES							
DATE	TIME	LOCATION	NUMBER	RATE	MIN	AMOUNT	RATE
	AMOUNT	TOTAL					
6/26	1123	SAN FRANCIS	415 555-1111	PEAK	2	.70	.08
	.78						
6/26	2210	INCOMING	415 555-2222	OFFP	3	.45	.00
	.45						
6/28	1348	OAKLAND	510 555-1111	PEAK	2	.00A	.08
	.08						
6/29	0941	LOS ANGELES	213 555-4444	PEAK	5	4.95R	.00
	4.95						
6/030	0832	WALNUT CK	510 555-3333	PEAK	2	.70	.08
	.78						

Calls identified with an "A" are free of airtime charges and are deducted from your airtime allowance.
Calls identified with an "R" are roamer calls in the city shown.

it. It might have an access charge of $29.95, with no free airtime included. Most people do not choose this plan because it does not include a discounted cellular telephone and offers no discounts on service for airtime volume or contract length. All of the other plans below assume that a one-year contract and a discounted telephone are included. The access charge for these plans would be lower for a two-year service agreement.

- The *economy* or *personal* plan is intended for personal or nonbusiness users. It has a low access fee, 30 minutes of airtime included free, high peak rates, and attractive off-peak rates. It might have a monthly access charge of about $19.95.

- The *business* or *executive plan* is intended for average business users. It has a higher access fee, 100 minutes of airtime included free, lower peak rates, and normal off-peak rates. It might have a monthly access charge of about $49.95. This plan might also include free features, like voice messaging, and/or some custom calling features, like Call Forwarding.

- The *volume* plan is intended for heavier business users. It might include 200 minutes of free airtime at an access charge of $79.95, with discounted prices for additional peak minutes of use.

Each of these plans has benefits for certain types of users. To illustrate, Table 5.1 shows how much each plan costs at various levels of use and mixes of peak and off-peak calling. The lowest-priced plan for each mix is shown at the bottom. Note that for these examples, each of the plans, except the basic plan, is the lowest priced plan for some mix of usage.

The Personal Plan provides the lowest total monthly bill at 50 minutes of use, regardless of whether usage is primarily during peak or off-peak times. One reason is because the access charge of the Business and Volume Plans pays for 100 or more minutes of use, which are wasted if actual usage is only 50 minutes. At 200 minutes of use, the Business Plan is cheaper than the Volume Plan only if usage is primarily off-peak. If usage is mainly peak, the extra discount on peak usage in the Volume Plan's included 200 minutes makes it a better value.

These plans are only illustrative, but show the relationships of the Plans for you to use to understand how to pick one. They also show that the differences between the costs of the various plans are significant, but not devastating. You should find that similar plans are available from your service providers.

ADDITIONAL RATE PLANS

Some additional rate plans for special segments of users are not included in the table, but may provide benefits to some classes of users.

- The *corporate* plan is intended for multiple users at a company, who agree to be serviced by a single bill. It might include 100 minutes of free airtime at an access charge of $69.95, with discounted prices for additional peak minutes of use.
- The *flex* plan automatically adjusts the airtime rates to be lower, depending on how much usage there is in a particular month. For users with widely varying usage from month to month, this provides usage discounts without requiring the user to change

Table 5.1 Comparison of Illustrative Rate Plans at Different Levels of Use

	Monthly Access	Included Minutes	Peak Usage	Off-Peak Usage	Monthly		Total Charges— Examples	
					50 Minutes		200 Minutes	
					80% peak	40% peak	80% peak	40% peak
Basic	29.95	0	0.40	0.25	48.45	45.45	103.95	91.95
Personal	19.95	30	0.75	0.25	32.95	28.95	130.45	96.45
Business	49.95	100	0.35	0.25	49.95	49.95	82.95	78.95
Volume	79.95	200	0.33	0.25	79.95	79.95	79.95	79.95
Lowest	Personal		Volume		Personal	Personal	Volume	Business

plans. Usage discounts are not as attractive as those for plans that require a fixed amount of use every month, but it provides economical usage flexibility for many business users.

- The *zone* plan provides a discount for use within a smaller geographic area than the entire system coverage area regardless of peak vs. off-peak usage, but charges a premium rate within the service area outside the zone of coverage. This is economical for personal users who use the phone at all times of day within a restricted area.

- The *prepaid* plan permits the user to have no contract and no monthly bill at all, and works just like prepaid long-distance cards. You purchase the phone at a higher price than with a contract, and purchase an amount of usage. You are informed at the beginning of every call of your current balance, and warned if it is low during the call. You replenish the account by calling Customer Service or stopping at an authorized outlet. Some vendors will let you use your choice of phones, and keep track of your bill via the network. Others require a special phone that keeps track of the charges itself.

  The prepaid plan is not just an option for those who can't pass the credit requirements for a contract, although it is very good for them. It is also recommended for those who want to budget or keep close track of their usage. Other applications are for teenagers when parents want to control costs but keep in touch.

- The *national plan* is a newer, innovative plan for heavy users who make many long distance calls or travel out of town a lot. It provides roaming at home rates anywhere in the country, with free long distance included in the airtime rate.

SPECIAL PROMOTIONS AND PRICING OPTIONS

By carefully shopping around you can make cellular more affordable. This applies not only to finding the best prices on equipment and accessories, but more importantly, to finding the best pricing plan for cellular service. The prices on cellular phones are so low that it is more important to select the right service plan than to run all over town to save $10 on a phone, only to find that the dealer does not service the carrier with which you want to subscribe.

Service providers offer special calling plans that provide additional options to their pricing plans at a fixed monthly rate. Such a plan might permit you to have free off-peak local calling, for example, or include long-distance charges.

Special promotions are limited-time offers run periodically by service providers or dealers that provide additional value. These may include service benefits of waiving the activation fee, free minutes of use, free voice messaging or off-peak calling for several months, or free voice messaging. Individual dealers or service providers may include free merchandise with the purchase of the combination of the cellular phone and subscription to the service. These might include an extra battery, carrying case, or complete starter kit of popular accessories, or a gift certificate for additional merchandise. Any of these promotions may be worth $50–$100 or more, and are designed to remove any lingering uncertainty you may have about finally signing up. They are worth looking for when shopping for cellular service and phones.

Finally, if you have a phone and wish to minimize your expenditures merely to have the phone available for emergencies, you can sign up with some carriers that offer a special plan that has no monthly access fee, but charges over $1.00 per minute for usage. Also, you may have to buy a phone if you don't have one—at a cost of $100–200, if you don't sign a contract.

HOW CALLS ARE BILLED

One thing that differentiates cellular charges from those of home and office phones is that whether you make a call or receive one, *you* pay for it (see Figure 5.2 on page 37).

When you place a call, the meter starts running when you press the SEND button on your phone to initiate contact with the cell site. Thus you are billed from the time you start to place your call, and not just for the time you are connected with the other party. In the world of cellular phones, *connect time* refers to the time you are connected (by radio) with the cell site and the mobile switching center (MSC), not with another phone. The calling period ends when you either hang up your mobile phone, or press the END button on your portable, or turn the phone off. Any of these actions terminate radio contact with the cell site. Some carriers charge for airtime at a reduced rate even if your call is not complete, because of the use of the radio channels to attempt the call.

When someone calls you at your cellular number, you pay for that, too (see Figure 5.3). The billing period starts when you answer and ends when you do any of the same actions as with an outgoing call.

Even though you are billed for outgoing calls as soon as the cell site is contacted, you only pay for completed calls, as with conventional phones. The reason the call timing starts when you press the SEND button is because you are using the cellular system's radio channel beginning at this time.

The reason that the airtime for incoming calls is billed to you is because there has been no agreement with conventional telephone companies to bill landline customers when they call cellular numbers. There is an issue that this *calling party pays* principle for cellular might cause concern to landline telephone customers if they call cellular numbers without knowing it is a cellular phone, and are charged airtime fees without prior notice on each call. Calling party pays has been initiated in a few cities, and it is hoped it will become standard in the future.

LONG DISTANCE AND ROAMING

From your cellular phone you can call—and receive calls from—anywhere in the country or in the world. The cost for a long-distance call is not much different from what you would be billed if you made the call from your home or office phone, assuming normal retail rates. In fact, the price is exactly the same, except that the normal cellular connect charges are added to it. Therefore, a call that would cost you, say, 29 cents a minute if you made it from your office, might cost you 69 cents a minute (29 cents plus 40 cents for airtime) when placed from your cellphone (see Figure 5.3). Remember also that some cellular service providers may also have an interconnection charge (for using the landline network to connect your call) of a few cents per minute, sometimes called, "landline charges" on your bill.

Many cellular service providers will permit you to choose your primary long-distance carrier, in the same way that you choose your

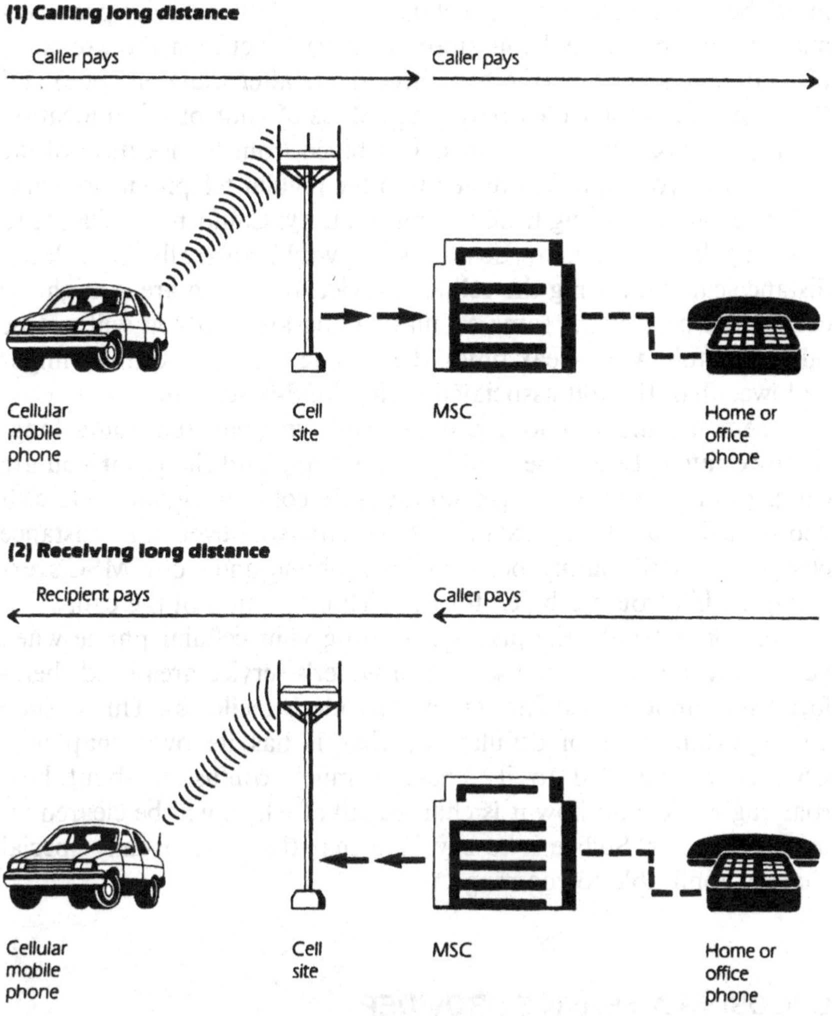

(1) Calling long distance

Caller pays

Caller pays

| Cellular mobile phone | Cell site | MSC | Home or office phone |

(2) Receiving long distance

Recipient pays

Caller pays

| Cellular mobile phone | Cell site | MSC | Home or office phone |

Figure 5.3   HOW A LONG DISTANCE CALL IS BILLED
[1] When you place a call, you pay the long-distance rate between the MSC and the point you are calling. [2] When someone calls you, they pay the rate between their phone and the MSC and you pay the portion of the call between the MSC and your cellular phone.

carrier with your home or office telephone service. The charges will appear on the cellular bill with the carrier identified.

In some areas, using your cellular phone will actually save you money, because a long-distance call from your office or home phone might be a local call using the cellular service. Other service providers may charge the normal toll amount for calls between the exchange of your cellular phone (the first three digits after the area code) and the exchange of the called party, regardless of your physical location when you make the call, because it is billed from the location of the MSC where your call is switched into the regular telephone network.

For example, calling from Orange County, California, adjacent to Los Angeles County, to Los Angeles would normally be a long-distance call, but, using the cellular service, these two areas might be considered part of the same cellular calling area, so you will not be billed any toll. At off-peak times, the airtime charge on cellular might be lower than the toll associated with a land-based call.

When you make a long-distance call, you pay the going long-distance rate between the MSC you are using and the point you are calling, plus the standard per-minute rate cost of cellular service. If people call *you* long distance, *they* pay whatever long-distance charges normally apply between their phone and your MSC's exchange, while you are billed for the airtime portion of the call.

Roaming refers to the practice of using your cellular phone when you're outside your home service provider's service area (and therefore using another cellular service provider's facilities). This is such an important part of cellular life that it has its own chapter — Chapter 7 — devoted to it. There is much confusion about how roaming works and how it is charged, all of which will be cleared up in that chapter. Suffice it to say here that there are usually special charges applicable to roaming.

CHOOSING A SERVICE PROVIDER

Since there are two cellular carriers in each area [and more carriers with the new personal communication service (PCS) services], you may want to consider the differences in service providers. In addition to the carriers, some cities have *resellers*, which buy service at wholesale from one or both of the two carriers and sell it at retail to end users, with their own rate plans, promotions, features, dealers, and customer service, which increases your choices. Service providers therefore include resellers and carriers.

Note that the dealer who sells you a cellular phone may have an exclusive arrangement with one service provider and may be able to offer service from that service provider only. You may like the customer service, installation, and phone you are offered by a dealer, but you might want to investigate service providers separately to make sure that you want to use the service provider associated with the dealer.

In addition to pricing differences (in both the structure and price levels of the rate plans) and periodic promotions or special incentives, service providers may be different in many other characteristics. One may have greater coverage in your area than the other does. One may have a larger area in which usage charges are toll-free.

Because systems can become congested with too many users, a service provider may have inferior service quality, characterized by dropped calls, several attempts required to put a call through, interference, static, and so on. Different providers also may have a different mix of special and premium services available.

Finally, customer service, billing convenience and simplicity, and other considerations can make the difference. The best way to evaluate service providers, if you don't have special needs in these areas, is to talk to current users. And remember our cellular buyer's checklist in Chapter 4.

6

HELLO, MA? IT'S ME!

Using a cellular telephone is not much different from using a conventional one. In fact, it can even be easier because these phones contain their own small computers and have many more special features than regular phones. And the high quality of their audio makes them a pleasure to use.

At first glance, the keypad of a cellular phone may appear intimidating. Once you know what everything's for, though, and how it works, there's nothing to it.

BUTTONS, BUTTONS

In addition to the usual twelve buttons found on a home or office Touch-Tone phone, the handset of a cellular phone contains several buttons that are intended specifically for cellular operation. Because convenience and safety are two prime factors in the design of a cellular phone, as many of its controls as possible — sometimes all of them — are grouped together on the handset of the vehicle-installed phone as well as on the portable.

The actual keys, key sequences, and indicators on any particular phone may be very different from those described here, because each manufacturer tries to make the phone's features more extensive and simpler to use at the same time. Your telephone's instruction manual and your service provider's user guide are your primary sources for information on the features available from the keypad and how to access them. This section provides general information on the use and representative functions of many of these keys for the prospective user. Figure 6.1 illustrates a typical portable telephone handset.

The PWR (power) button (8) turns the phone on and off. Often this button must be depressed for a fraction of a second longer than you expect for it to work; this is done to prevent accidental or inadvertent power off/power on.

Remember that the phone must be turned on in order to receive calls as well as place them. Keep the phone on when you expect a call. Turn it off when you are not available for calls or when it is unsafe to use. It is against Federal Aviation Administration (FAA) regulations, for example, to operate a cellular phone in an airplane. And etiquette requires that your phone shouldn't ring, nor should you take calls, at a concert or a business meeting. Use the PWR button to control your availability. At your home or office, a ringing phone demands answering, but you don't have to be available for calls all the time. Calls can be diverted using Call Forwarding, or sent to a voice messaging service.

When the power is turned on, you may be surprised by the flashing of the display and the tones emitted by the phone. This is merely the phone going through a self-test diagnostic, to make sure it is operating correctly.

The SND (send) key (1) puts the phone into action when you are ready to place a call. Until it is pressed, no call will be initiated or received; nothing will go over the air or into the phone system. This is the main difference between using a cellular phone and a conventional phone, where a call is initiated automatically as soon as you are finished dialing the number. In this way, the radio channel is not used during the dialing procedure, saving valuable radio capacity.

This procedure of pressing the SND key to initiate a call allows you to dial a number at any pace you choose while parked or at a stop light in your car, check it on the display, and then start a call with one touch of the SND button when it is convenient. You can still receive calls while dialing in this fashion.

When the number to be called is available in the phone's working memory (discussed below), you can press the SND button and you'll be on the air. The IN USE indicator will light to show you that a connection with the cell site has been established and your call is in

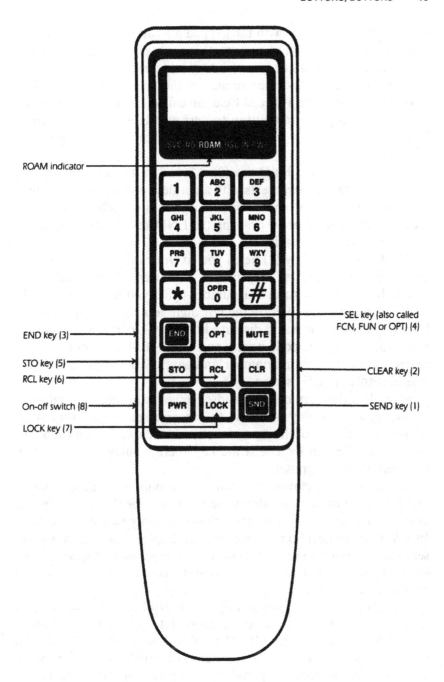

Figure 6.1 THE KEYPAD OF A CELLULAR PHONE

process. You can fiddle with the keypad all you like at any time, but until you press the SND key, everything you do will affect only you and your phone, and not go out into the system.

The SND key is also used to answer the phone. Many models of portable phones have a special function called *any-key answer*, which permits you to press any button to answer the phone when it rings. Car phones also allow you to answer the phone by picking up the handset.

The purpose of the END key (3) is found in its name. It's used to hang up the phone, because as with a cordless phone, there's nothing to "hang up" on a portable phone. Car phones will allow you to end a call by hanging up or by pressing the END key, when operating in hands-free mode, for example.

The CLR (clear) key (2) helps you fix your mistakes. It is all too easy to hit the wong key, especially if you disregard the advice given in this book and try to dial a number while you're in motion. The CLR key allows you to erase the last digit you entered, or if pressed longer, will erase the entire displayed number—perhaps you got the first digit of the area code wrong and didn't realize it until you had entered the entire phone number and saw it on the phone's liquid crystal display (LCD) screen.

Several of the keys on the handset serve more than one purpose. It is designed this way primarily to conserve space on a very small telephone—if each of the many functions available from a cellphone had to be called up by a separate key, there wouldn't be any room for them all on a handset.

These extra functions may vary from phone to phone, but a number are found on just about every phone. For those with special keys, most functions require the pressing of only two or three keys. In addition, some phones have a large display, and allow you to select the setting for special functions using a menu and special menu keys. On some phones, the display can also be used to send you messages or to act as a pager.

For those phones with multiple key strokes for special functions, the first key that must be pressed to set such functions is the FCN (function) key (4). The FCN key is pressed, followed by a second key. It may be labeled SEL, OPT, or FUN. Some examples, for Motorola's popular flip-phone portable models, are the following:

FCN-#   Displays the telephone number of your telephone
FCN-G   Displays the battery strength
FCN-S   Displays the signal strength

The STO (store) key (5) allows you to store numbers in the phone's memory. On some phones, calling-card numbers and other numbers also can be recalled from memory, and can be up to thirty-two digits long. Some phones have ninety-nine or more memory locations, while most people use less than five. To store a number in memory on most phones, enter it in the display, then press STO and the two-digit number of the memory location desired. Some phones may require that you press the STO key again to confirm storing the number.

To recall the number from memory, the RCL (recall) key (6) is used, followed by the number of the memory location of the number desired. Most people put their office and home-phone numbers in the first two locations. Many phones allow you to associate a name with the stored number and recall the number by entering the name. To place a call, merely press the SND button after recalling the number to the display. Make sure the saved number includes a "1" prefix or area code if required. If your carrier will accept calls to local numbers preceded by the local area code, it is easiest to record all numbers with an area code, so that the number can be used locally or when roaming, without having to modify the stored number. Some phones allow you to re-call the last number dialed by pressing the RCL key only, or pressing the sequence RCL-#.

The LOCK key (7) is part of a phone's security system. When you press it, the word LOCK will appear on the display, and it will be impossible to dial out, although the phone will still be able to receive calls, and perhaps dial 911. This feature is designed to protect you from unauthorized use, or in case the phone is stolen.

To unlock the phone, you must enter a sequence of numbers, perhaps only four or five of them, known only to you. It may be your birth date, part of your home-phone number, or any other figure that is easy to remember. Never write down your unlock code where it can be easily found and associated with your phone—it might be discovered by someone who has no business having it and who might be tempted to use your phone without your knowledge or permission.

On car phones, the HORN button, when activated, may be configured to use your horn to signal you for an incoming call, and allows your phone to remain on with the ignition off.

The MUTE key allows you to turn off the microphone on your phone temporarily for private discussions on your end.

Some type of volume control is usually included, either via a key, or using a rotating dial or a toggle switch on the side of the unit. Many phones permit you to change the sound and/or the volume of the ring.

DISPLAYS AND INDICATORS

Various indicators and displays on the phone tell you a lot about its current status. Some phones have a limited display that may not even display an entire ten-digit telephone number, may use difficult-to-read alphabetic characters, and may use small light-emitting diodes (LEDs) for indicators. Most new phones have a larger display using an LCD, which includes both numbers and letters, as well as graphical displays of status indicators. It is important that you can see the display easily in all kinds of light. Some of the most often used indicators are described here.

Of course, the power indicator (PWR) tells you when the phone is on and ready to take or make calls. The IN USE indicator tells you that you are in contact with a cell site, and that your call is being processed or in progress. If you check this indicator when encountering interference, it will tell you whether or not you have lost a call. The NO SERVICE indicator lights when you do not have coverage or have inadequate signal strength to make or receive a call. The CALL indicator tells you that a call was received but not answered.

The ROAM indicator tells you that you are accessing the system of another carrier than your own. When traveling outside your home service area, this light will light. If you are in your home service area and your own carrier's signal is poor, the phone will try to access the second carrier's system, if your phone is set to access either system. In this case, since you are changing from block-A frequencies to block-B frequencies, or vice versa, the ROAM light will *blink*. This will also occur if you access the opposite carrier from your home carrier's block when you travel to another set of two systems. When you are accessing the other carrier's system in this way in your home area, you will usually be connected to the other carrier's customer service if you try to make a call, since you are not a subscriber to their system, and you will not receive incoming calls. Cellular operators in the same area do not normally cooperate on these matters, because of competitive concerns.

Many cellular telephones have additional display indicators. Some have the battery level and signal strength always displayed graphically whenever the phone is on.

PLACING A CALL

When you pick up the handset of a cellular phone, you don't hear a dial tone. That's because, until you press the SND key, you aren't connected with the cell site (where the signal originates) and also because, in cellular telephony, a dial tone isn't necessary. The purpose of the dial tone is to tell you that you are connected with the phone system's switching equipment; a cellular system has a different way of doing that.

You dial a number just as you would from a standard phone, by pushing the buttons corresponding to the digits. As you enter the digits, they appear on the display above the keypad (Figure 6.1), each digit being bumped along from right to left as new ones are keyed in (You can even dial a number the night before if you feel like it.) These digits will stay in the phone's working memory until you clear them out or until they are replaced by another number.

If the number you want is already stored in the phone's memory, simply press the phone's RCL button, followed by the appropriate code. If you've forgotten the code—usually a number from 01 to 99 in the case of phones that can store 99 numbers—there is usually a way for you to scan through the codes and numbers to see which goes with which. As stated previously, many phones have names associated with the numbers in memory; others allow you to avoid the RCL button altogether and scan the names of people or numbers stored until the desired one is displayed.

With the number now in the phone's working memory (indicated by its being shown in the display), you are ready to establish the connection by pressing the SND button. This activates the phone's transmitter and sends the dialing information to the nearest (or strongest) cell site, from which it will be sent on to the mobile switching center (MSC). If you are successful in reaching the cell site—as you usually will be—the phone's IN USE indicator lights up. If, for some reason, your phone is unsuccessful in completing the call, you should receive an audio indication from the phone. If you are unable to reach a cell site, this usually sounds like a warbling busy signal. Probably, your NO SERVICE indicator is also illuminated. A nonwarbling fast busy signal tells you that you got through to the cell site, but the system is busy. Sometimes the reason you can't complete a call may not be so obvious. You may not be a subscriber to the system, or you may have dialed incorrectly.

If your phone's NO SERVICE indicator is on, it usually means that you are not within range of a cell site. This may be the case even if the NO SERVICE indicator is not lit when you check it. The reasons for this happening and the ways in which you can improve your chances of getting through under difficult conditions are detailed in Chapter 10, "Dealing with Operational Difficulties." You'll find that the conditions under which you can receive a call or maintain a connection in a "spotty" area are different from those required to place a call. After a little practice, you will learn to tell how long you should expect to wait for a connection to be established and what to do to improve your chances under less-than-ideal conditions.

If you are unsuccessful in establishing a connection the first time, try again by pressing the SND button—the number is still in the phone's working memory even if it has disappeared from the display (it will reappear when you press the SND button).

Unlike conventional phone systems, as your call progresses toward the point where you are connected with the number you've dialed, a cellular system lets you hear no tones, clicks, or beeps. What you will hear from your phone's speaker is a quiet hum, which is the carrier signal from the cell site. (It is like the sound you hear when a broadcast radio station is on the air but is not transmitting any speech or music.)

The first sound you'll hear from your phone will be one indicating that the remote phone is ringing...or that it is busy. In the normal course of events, the next thing you'll hear will be that of the remote handset being picked up and a welcome "Hello."

To terminate a cellular phone conversation, merely press the END key. On car and transportables that have a separate handset and cradle, you may hang up the handset to end a call, as with a conventional phone. These phones also allow you to use the END button to end a call. This is convenient, for example, when you want to immediately place another call while holding the handset.

Remember that if you don't terminate the call, even though the party on the other end has hung up—whether you placed the call or just answered it—you're still connected to the cell site and paying for that connect time for the next 20 to 30 seconds. After that, the MSC will usually disconnect your call.

SPECIAL TELEPHONE NUMBERS

There are special services available from your carrier, many of which have abbreviated dialing codes, often prefixed by the asterisk "*" on the dial, called "star." Thus, "STAR-6-1-1" (or *611) will usually connect you to your carrier's customer service department. If you dial "611" without the "*," you would get the local telephone company's customer service, as you would if you dialed "611" from your home or office.

There are additional services available from many carriers. In some areas, dialing "*99" might get you into your voice-messaging system to retrieve messages. "*JAM" might get you to traffic information. Some of these services may have airtime charges plus additional usage charges; customer service calls may have no charge at all. "*411," for example, is not the standard phone company directory assistance, but a premium directory assistance service offered by some carriers. In addition to finding listings, this service can recommend a restaurant or find a taxi for you, and for an additional charge, perhaps dial the number for you. If you dial only "411," without the star, you will get the local telephone company's directory assistance services.

It is important to remember to use "911" without the star, as with traditional telephone services, for emergencies, where it is available. With cellular, however, the 911 dispatcher will not be able to automatically receive the location of the phone as it does with landline service. You need to report your location. Cellular has greatly enhanced safety because it makes telephone service much more available for emergencies away from traditional telephone service.

RECEIVING A CALL

Receiving a call on a cellular system is even easier than placing one, always assuming that your phone is turned on and that you're within range of a cell site, which is indicated by the fact that the NO SERVICE light is not on.

With a cellular phone, you're never far from important calls. (Photo courtesy of Motorola, Inc.)

When someone dials the number of your cellular phone and is connected, your phone will ring (on some phones you may have a vibrating option on your phone that you can activate, for example, if you're at a concert), and callers will hear a ring signal at their end. There is no bell on a cellphone—the ringing sound is generated electronically and reproduced by the phone's speaker. All you have to do is press the SND button (on some phones, pressing *any* key will work). On a car phone or transportable, you can pick up the handset to answer, or press the SND button to use a hands-free system. Don't forget that it's up to you to hang up!

The IN USE indicator will light whenever you have a call in progress, during call processing or conversation. When you end a call, the IN USE light will go out to confirm the call has been terminated.

Some users never learn more than the SND and END buttons, and get a lot of value with only this simple operation of the phone and cellular service. Cellular's simplicity is one of the keys to its popularity. But there are many additional features, conveniences, and services available that are literally in the palm of your hand with a cellular phone, if you care to take advantage of them. We've covered a few, like re-calling numbers from memory, but we will discuss more below.

HEALTH CONSIDERATIONS

From time to time rumors have surfaced that there may be some health risks to the use of cellular phones, especially portables, due to radiation from the tip of the antenna in close proximity to the head. While studies continue and more stringent standards on radiation are being considered, it should be pointed out that no studies have linked any cellular phones to health risks. Any assertions of health risks up to this point are merely rumors and uninformed accusations.

BATTERIES AND POWER SUPPLIES

Batteries are an important subject for portable-phone users. Infrequent or occasional users want to make sure the battery will not go dead while the phone sits unused. Heavy users want the battery to last as long as possible between charges. Whether you use a car-installed mobile phone or a portable, a cellphone operates from one type of rechargeable battery or another. A cellular phone can operate from its own battery or a car battery.

The information that follows will help you choose the best battery, and extend the life of the batteries that keep your cell phone operating.

Battery Types

Portable cellular phones rely on three types of rechargeable batteries: nickel–cadmium, nickel–metal-hydride, and lithium-ion.

Nickel–cadmium batteries, or NiCads, are the standard type of rechargeable battery used in such household appliances as cordless vacuum cleaners and flashlights. They suffer from the "memory" effect, developing a memory for being used a certain amount of time before recharge, and only charging that much when recharged. If you use a phone powered by NiCad cells just a few times before recharging it, instead of running it until its batteries are nearly exhausted, it will begin to demand recharging before it is necessary, and retain only a small charge. You should charge NiCads before using them, and use them until the battery indicator shows their charge is nearly completely depleted. If NiCad batteries develop a memory problem, you can often restore them by running them through two or three complete charge/discharge cycles.

Nickel–metal–hydride (NiMH) batteries do not exhibit memory effects, and hold a charge longer than nickel–cadmium batteries. Lithium-ion batteries are the newest rechargeable battery type, and show even greater capacity for their size. This permits longer talk and standby times between charges. But NiMH and lithium-ion batteries are also more expensive than NiCad batteries.

CELLULAR ETIQUETTE

The general availability of portable cellular telephones that can easily be carried in the purse, briefcase, or suit/shirt pocket has generated a number of new social situations inconceivable before the advent of the portable cellphone. The results are sometimes humorous, but more often embarrassing and annoying to others.

Your new ability to be in constant communications (or merely to impress people) can make you oblivious to your potential to infringe on the rights of others when using a cellular phone. Try to be courteous in the following situations:

1. Don't put your cellular phone out on the table at a business meeting or at a restaurant; this sends a signal that you may interrupt the meeting or the meal with a phone call, and that you value the phone call more than your present company, business or personal.

2. Don't use a cellphone at a meeting, even if at a break; excuse yourself to another room or the corridor.

3. Don't use the phone at a table at a restaurant; it is discourteous to other patrons as well as those at your table; excuse yourself to the lounge.

4. Don't let your portable phone ring at a theater, meeting, or conference, much less answer it or make a call. If you must monitor incoming calls, use a pager with silent notification. If you are expecting an important call, reschedule it.

5. Don't ask others at a gathering to stop their activities, to be quiet, or to leave the room so you can hear your telephone conversation or make a call. Your new ability to make or take calls at any time does not preempt the activities of others.

6. Use discretion when interrupting any conversation or activity with others to make or receive a call. A good rule of thumb is whether you would interrupt such a conversation for a conventional telephone call. Excuse yourself, and do it *before* you answer the phone—don't carry on even the beginning of the telephone conversation in front of another.

While these are important guidelines for courtesy, some people just love to be noticed using a cellular phone—and violating these guidelines can certainly make you the center of attention!

SPECIAL CELLULAR FEATURES

Companies that provide cellular phone service offer their customers a number of optional features that can be added to the basic service.

These features can tailor the service to your particular requirements. Some of these features were mentioned in the previous chapter under "Recurring Expenses"; a more comprehensive list follows. Generally, these features are available at a monthly cost of about $3 apiece. Some high-end subscriber pricing plans for heavier users may include several of these features at no charge.

It may become difficult to remember which features are operated within the phone, and which are being performed by the cellular carrier, since both are controlled by keystrokes on the phone. Telephone number memory, or Speed Calling, for example, is in every cellphone, but is also offered by many carriers as a feature to pay for every month.

Local Number Plan Area Calling Only

Local Number Plan Area Calling Only restricts calling to the local service area, so that no long-distance toll calls are placed from the cellphone. The local service area is not necessarily a single area code or exchange, because the cellular coverage area does not usually coincide with the telephone exchange or area code geographic boundaries.

If your phone is used by a number of people, or if you are responsible for a fleet of phones used by your company's sales force, this feature can save money by making it impossible to use the phone for long-distance calls billed to the phone. Of course, calling local numbers to place collect and calling-card long-distance calls is permitted. If you anticipate that all the calls from your cellular phone should be local ones (incoming long-distance calls are not affected), the slight monthly charge can be a small price to pay to guarantee that the phone won't be misused.

Speed Calling

Speed Calling allows you to store a list of eight or more frequently called numbers at the mobile switching center (MSC) instead of in the phone, and to call them by dialing one or two digits. Now that most cellular phones include a large telephone number memory, this feature is somewhat superfluous.

Call Forwarding

If you're expecting an important call at your cellular number but can't stay by the phone, Call Forwarding allows you, by pressing a few keys on the cellular phone, to have calls placed to your number automatically transferred to any other number. Thus, incoming calls can be referred to a voice-messaging system or assistant. In this way, incoming-call airtime charges can be avoided, or you can have messages taken to avoid missed calls. You must pay for the additional toll from your cellular number's exchange to the exchange of the final destination, if applicable, and usage charges may apply in some cases.

Conditional Call Forwarding

Conditional Call Forwarding works like Call Forwarding, but calls are only transferred if you do not answer your telephone after several rings, or the line is busy. This permits you to take calls when available, or leave them to be transferred to voice messaging or an assistant. *No-Answer Transfer* is a version of this feature for transferring the call only when it is not answered.

Call Waiting

Call Waiting gives you the equivalent of a multiline phone at your cellular number, allowing you to be reached even when you are already using your phone.

Just like the landline version, if people try to call your cellular number while it is in use, instead of a busy signal they will hear a normal ringing sound. Your phone beeps to alert you to a second caller. You can end the first conversation to begin the second, or you can alternate between the two. Note that you will be billed for both calls. Only people who are very heavy users and expect a lot of important incoming calls that must be answered in real time will find this service useful for cellular. Conditional Call Forwarding will serve most users by forwarding the second call to a voice-messaging system.

Three-Party Conferencing

Three-Party Conferencing allows a third party, from a third number, to be added to a conversation. You are billed for connect time to both other numbers.

Voice Messaging

Voice messaging is usually used for *telephone answering* on cellular, instead of having a telephone answering machine. (Cellular telephones are available that also have a built-in answering machine capability.) Many cellular users also have voice messaging available at their office, and forward calls to their office to take messages.

Voice messaging is usually offered as a special service by the cellular service provider, through a specialized computer attached to the MSC. In conjunction with Call Forwarding or Conditional Call Forwarding, it will answer your phone when you are not available or don't care to answer. Some of these services can call you or page you to tell you that you have an important message waiting, or permit you to exchange long messages with other users who are business associates without calling them directly; this is called *voice mail*.

Voice messaging is the general term for this capability, which includes *telephone answering* and *voice mail* as applications. Some of these services allow you to receive, store, and send fax messages. This permits you to have faxes sent to you at any time, reroute them to a fax machine convenient to your next destination, or print them all on your portable fax machine at a convenient time. Or, you can record a message or fax in your mailbox, and "broadcast" it to an entire group without having to send it to each person individually.

Voice Dialing

Voice Dialing permits you to use your voice to dial the phone. You can say the digits, or just say a name associated with the number that you've previously programmed, like "Pat" or "Office." This is helpful when you've got a lot of numbers stored. It's also less distracting when you're in a vehicle.

7

ROAMING

Using your cellular phone when you're away from your home cellular service area, traveling in another, is called *roaming*. Virtually all cellular carriers have made arrangements under which you can simply arrive in a service area other than your own and automatically start to use that carrier's service.

There is a certain mystery surrounding roaming, and a lot of questions have arisen over what roaming is and how it works. This chapter is intended to answer those questions.

HOW TO TELL WHEN YOU'RE ROAMING

The boundaries that mark the end of one cellular carrier's domain and the beginning of another's are invisible. Unless you've traveled the route before, you can't tell when you've left your home area and entered another. But your phone can.

One of the indicators on a cellular phone is labeled ROAM (see Figure 6.1). It serves several purposes. When you leave your home region and enter another, the ROAM indicator will light and stay lit as long as you are within the range of a cell site. If you are between systems or out of range of the cell site of any system, the ROAM indicator will go out and be replaced by the NO SERVICE light as you pass through the area.

The ROAM indicator serves another purpose. It will tell you when you are receiving signal from a carrier on a different frequency block from your home carrier. All cellphones are equipped to switch automatically between two services, nonwireline and wireline, or the A vs. B block carrier. The service you normally use is programmed in the phone's numeric assignment module (NAM) to give one priority over the other, so that your own carrier will always be selected over the competitor in your area. On many phones, you can program which carrier, A or B, is to have priority over the other yourself, and even lock out the other completely. The mechanism that does this is called an A-B switch. All cellular phones have A-B switches, but some are more versatile than others.

If you leave your A-B switch in automatic or priority mode as you travel, your phone's ROAM indicator, instead of lighting steadily when you enter another service area, may begin to flash slowly. This tells you that the service your phone is "listening to" is a service other than the one it has programmed as priority — a B carrier when you normally use an A carrier, or vice versa. Usually your phone will automatically make the switchover, and, except for the flashing light, there will be no difference apparent to you.

Some carriers operate many adjoining systems on the same frequency block, with special pricing for roaming, so you may want to use your phone when roaming only when the ROAM light is on but not flashing, indicating you are receiving signal on the same frequency block as your home carrier. When you first enter a new area, you may first receive signal from either carrier. As you move throughout the new area, your ROAM light should stop blinking as you receive signal from both carriers and your priority carrier is chosen. Some carriers operate different systems on opposite frequency blocks. In this case, you have to switch to the opposite frequency block to stay with the same carrier.

The flashing indicator can be startling the first time you notice it. Don't be alarmed, there's nothing wrong with your phone. It's just keeping you informed of what's going on.

ROAMING: OUTGOING CALLS

If it weren't for the ROAM indicator and the changing scenery, you might never know that you were away from home, calling on a service other than the one you normally use.

If you roam to a system with a different area code, all the numbers you normally dial as a seven-digit number in your home area will have to have the appropriate area code first. This is especially difficult to remember when using numbers from your phone's memory. As long as your carrier allows you to include the local area code when dialing local numbers, it is best to enter all telephone numbers in your phone's memory with the area code, so that they can be used anywhere.

Some cellular services have adopted a policy of requiring an area code for all calls, even local ones. This eliminates any possibility of confusion. If you omit this where it's required, you'll get a recording telling you to dial again.

You may be accustomed to dialing the numeral "1" before an area code when calling long distance from a conventional home or office phone. Some cellular carriers do not require this. All you need is the three-digit area code, followed by the additional seven digits specifying the exchange and number. Other carriers may require you to use the leading "1," so if you forget to include it, you may get a recording requesting you to dial again with the leading "1" first.

As mentioned, many service providers have special abbreviated codes for traffic information, customer service, retrieving voice messages, and so on. Remember that these codes may change from system to system.

ROAMING: INCOMING CALLS

If you're within range of a cellular system, no matter where you started out from or where you may be, you can receive calls at your cellular number. It's a bit more complicated, however, than placing them.

The problem is that callers may have to know what area you are roaming in. There are two methods of receiving incoming calls while roaming. The first is automatic, and not uniformly implemented by all carriers. The second is a more cumbersome, manual method.

If someone calls you at your regular cellular number and you are roaming in another cellular system, the home system may forward your calls to your phone if the mobile switching center (MSC) knows which system you're on. This is called "follow-me" roaming, or *automatic call delivery*.

There are several methods to register your presence in a foreign system (a system other than your home system) so this will work. Some systems require you to enter a short code and press the SND button. Others will register you automatically when you place your first outgoing call. These alternative actions register you on the Visiting Location Register (VLR) of the visited system's MSC. The home system has you registered as a regular customer on the Home Location Register (HLR), part of its MSC. The visited system notifies your home system of your location, and calls to your cellular number will be forwarded to your phone in the foreign system. If the call is unanswered, however, your call may not be forwarded to your voice-messaging system or other number where you might ordinarily forward unanswered calls when you are in your home area.

The second, more manual method, is for callers to call a *roamer access number* in the city in which you are roaming. They will receive a second dial tone, inviting them to then dial your regular cellular number, including the normal, home area code. The foreign system can then see if a phone with that number is present on its system, as your phone is constantly communicating with any system when it's on, identifying itself to the system by electronic serial number (ESN) and home system telephone number, contained in its NAM. If it locates you, it can connect the call.

Although call delivery is the easiest method to receive incoming phone calls while roaming, there are several reasons why you might want people to call you using the roamer access number. First, the caller will pay the long-distance charges to your location from their location, instead of you paying long-distance charges from your home city to the city where you are roaming under automatic call delivery. Second, if the caller is located in the same city as you, neither of you will pay long-distance charges, where you both would if the caller paid long-distance charges to call your cellular number in your home city, and the call was automatically forwarded to you at your location at your expense. Third, sometimes you need to register as a roamer for automatic call delivery, and you may forget, or the call-delivery feature may be out of service.

The roamer access number is a regular telephone number composed of the visited system's area code, exchange, and number — usually R-O-A-M (7-6-2-6). The caller needs to know this number in advance. Your own carrier's customer service department can tell you the roamer access number for the city you are visiting and the carrier you plan to use, so you can tell callers in advance. Some roamer access numbers are listed in Table 7.1.

ROAMING FRAUD

Several different schemes have been devised to attempt to obtain free cellular service. This activity is unlawful, and carriers are actively seeking to help law enforcement capture and punish those who practice it.

In some cases, cellphones are programmed with fictional telephone numbers from foreign cities that cannot be easily verified by the foreign carrier, and rapidly changed in repeated dialing attempts until the carrier accepts the call. This is called *tumbling*. A more sophisticated scheme, called *cloning*, involves the interception of the ESN and telephone number of legitimate users from the airwaves, and "cloning" them into another cellular phone. When a criminal

Table 7.1 Roamer Access Numbers

City	Carrier	Roamer Access Number (7626 = R-O-A-M)
Atlanta	A	404 558-7626
	B	404 372-7626
Baltimore	A	410 208-7626
	B	410 382-7626
Boston	A	617 633-7626
	B	617 285-7626
Chicago	A	312 659-7626
	B	312 550-7626
Dallas	A	214 850-7626
	B	214 384-7626
Denver	A	303 888-7626
	B	303 877-7626
Detroit	A	313 938-7626
	B	313 320-7626
Houston	A	713 825-7626
	B	713 824-7626
Indianapolis	A	317 443-7626
	B	317 432-7626
Kansas City	A	816 591-7626
	B	816 223-7626
Las Vegas	A	702 595-7626
	B	702 379-7626
Los Angeles	A	213 712-7626
	B	213 718-7626
Miami	A	305 794-7626
	B	305 343-7626
Milwaukee	A	414 254-7626
	B	414 791-7626
Minneapolis	A	612 867-7626
	B	612 720-7626
New Orleans	A	504 583-7626
	B	504 450-7626
New York	A	917 847-7626
	B	917 301-7626
Philadelphia	A	215 350-7626
	B	215 870-7626
Pittsburgh	A	412 298-7626
	B	412 855-7626
St. Louis	A	314 973-7626
	B	314 277-7626

uses a phone number from one city and "clones" it into a phone in a second city, the carrier in the second city thinks it is a roamer. This is called *roaming fraud* and makes fraud criminals harder to track.

But as these schemes have been uncovered, new methods of securing the cellular system and checking the offered combinations against the home data base in real time have been developed.

Fraud has unfortunately inconvenienced many legitimate cellular users. Many systems sometimes require special codes, or personal identification numbers (PINs), like automatic-teller-machine (ATM) passwords, to be appended to users' calls when dialing. Roaming has even been temporarily discontinued in some cities to combat roaming fraud. And legitimate subscribers whose telephone numbers are used for such fraud must cancel and change their cellular telephone number in order to stop a cloned phone. Thus, you may find that roaming is not available in some cities, or requires you to get a special PIN code in order to make calls home or away.

ROAMER BILLING

With all this calling back and forth going on, you might be confused about who pays for what in a roaming situation. It's not difficult to sort out if you look at things one piece at a time.

A Roamer Calls Within a Foreign Carrier's Calling Area

This, of course, is the most straightforward case, where a roamer makes a call to a number located within the area he or she is visiting, which can be done by merely dialing the number (including the area code).

Each cellular system has its own rates for roaming callers (the charges will appear on your regular statement). Roaming charges vary considerably from system to system, as do normal cellular rates. Several of the larger carriers cover entire regions of the country, and offer reduced rates for their subscribers roaming onto their system in another city. Many carriers charge a daily rate of $3 or so for access. Your charges are forwarded to your home carrier to include in your bill, and sometimes may not appear on the next bill because of the delay caused by transferring the information. Table 7.2 shows the rates charged in several cities. The cost per minute to roamers is

Table 7.2 Roaming Rates for Several Major Cities

City	Carrier	Typical Rates Per Day	Typical Rates Per Min.
Atlanta	A		$0.60
	B		
			$0.65
Baltimore	A	$3.00	$0.95
	B		$0.65
Boston	A	$3.00	$0.99
	B		$1.09
Chicago	A	$3.00	$0.85
	B		$0.65
Dallas	A		$0.99
	B		$0.65
Denver	A		$0.99
	B		$0.65
Detroit	A		$0.99
	B		$0.65
Houston	A	$3.00	$0.99
	B		$0.65
Indianapolis	A	$3.00	$0.85
	B		$0.65
Kansas City	A		$0.99
	B		$0.65
Las Vegas	A	$3,00	$0.99
	B	$3.00	$1.09
Los Angeles	A		$0.99
	B	$2.00	$1.09
Miami	A		$0.99
	B		$0.65
Milwaukee	A		$0.65
	B		$0.65
Minneapolis	A		$0.99
	B		$0.65
New Orleans	A	$3.00	$0.99
	B		$0.65
New York	A		$0.99
	B		$1.09
Philadelphia	A	$3.00	$0.99
	B		$0.65
Pittsburgh	A		$0.99
	B		$0.65
St. Louis	A	$3.00	$0.99
	B		$0.65

usually somewhat higher than it is to local users because of the extra costs involved in billing, and the lack of monthly access charges. These per-minute charges are applicable whether you place or receive a call as a roamer. A few systems charge for incomplete calls — a busy or no answer — because the channel is used whether the call is completed or not.

A Roamer Calls Home

When roamers place a call to their home calling area, they are billed the normal (roamer's) per-minute usage charge of the area they are in as well as the long-distance charges between the exchange of the MSC they are using and the exchange they are calling.

A Roamer Calls Someplace Other Than the Home Area

When roamers place a call to someplace other than their home area, they of course have to pay the per-minute airtime charges. They are also billed the long-distance rate between the MSC of the visited system and the place called.

A Roamer Gets a Long-Distance Call With a Roamer Access Number

When someone calls from a landline phone in the home area using a roamer access number to reach a roamer, the roamer is billed per minute at the standard roamer rate for airtime. The long-distance charges are billed to the calling party.

A Roamer Gets a Long-Distance Call Without a Roamer Access Number

A call from a landline phone directly to the subscriber's number goes first to the home system MSC, then is forwarded to the roamer in the visited system. Again, roamers pay the airtime charges between their phone and the exchange of the MSC they are using. But the caller pays the long-distance charges only to the home MSC. The roamer pays the additional long-distance cost to the visiting MSC from the home MSC.

ADDITIONAL ROAMING CONSIDERATIONS

This is the way things work for most areas. In some parts of the country, however, you must write or call ahead to make arrangements to use your phone during a visit. Sometimes roaming can even be temporarily prohibited by the carrier, because of roaming fraud. Before you go anywhere you should check with the carriers you plan to use to determine their requirements. Your own cellular service's customer service operation can frequently assist you in arranging to use another facility, including advising you about their requirements, roamer access number, and registration method.

Most cellular carriers have agreements that make prearrangements for roaming unnecessary. If you do need to make arrangements, you may need to have your telephone's ESN and telephone number ready when you call. Some carriers accept credit cards for roamer service.

If for some reason regular roamer service is unavailable to you in a city, you can sometimes sign up with a carrier to use their system using a special plan that has no monthly access fee, but charges over $1.00 per minute for usage. In chapter 5 we mentioned this as a rate plan for those who have a phone; if you have a credit card, this can be a useful alternative if normal roaming has been suspended in a city due to fraud problems.

If you use cellular service in another carrier's service area often, you may want to subscribe as a home user in that city. Many cellular phones have dual NAMs that allow you to register as a home subscriber in a second city and pay the monthly access charge, so you pay a premium to enjoy regular subscribers' usage rates instead of roaming rates. You will have a separate local number in that city.

You should know that your Advanced Mobile Phone Service (AMPS) cellular phone will only work in the North America, including Canada. For other countries, other cellular standards and frequencies are used in most cases. You can rent a cellular phone for use in many countries. Some new multimode, multiband phones are appearing for use in multiple countries, including the United States.

While we've tried to cover a lot of local variations, for most cellular users roaming should be fairly routine. In the near future, arrangements for roaming and the forwarding of calls should become automatic and standardized. If you are having trouble as a roamer, you can usually get through to the local carrier's customer service by dialing *611 for help. Some carriers will call *you* when you first enter a cellular system if your phone is turned on, because the MSC will sense your presence on the system. They will welcome you as a roamer and inform you how to get more information on using the system — and, of course, they won't charge you for this call.

8

MOBILE AND TRANSPORTABLE PHONES

To be able to place and receive telephone calls for various wireless applications—a telephone for the construction site, or on a boat, for example—you may want to consider a transportable phone. Vehicle-installed phones used to be the primary type of cellular phone, and are still important for many users.

CAR CELLULAR TELEPHONE COMPONENTS

A typical cellular phone installation for an automobile consists of four basic components: the power source, the control head, the transceiver/logic unit, and the antenna (see Figure 8.1). Most problems with installed car phones are a result of poor or improper installation, resulting from the need to connect several components on many different vehicle models. These components are incorporated into a single unit in the transportable and portable cellphone, but should be considered logical parts of any cellular phone. Many vehicle-installed phones have some of these components incorporated into a single multi-functional component.

Power Source

In most cellular phone installations the phone receives its power from an automobile battery. If it is part of a permanent installation, the

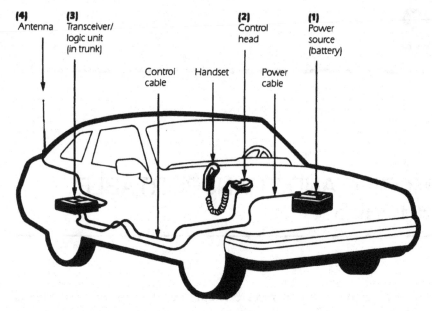

(4)
Antenna

(3)
Transceiver/
logic unit
(in trunk)

Control
cable

Handset

Power
cable

(2)
Control
head

(1)
Power
source
(battery)

Figure 8.1 A TYPICAL INSTALLATION
*A typical cellular phone installation consists of four basic components: [1]
power source, [2] control head, [3] transceiver/logic unit, and [4] antenna.*

phone may be permanently connected, or hardwired, to the vehicle's
electrical system. In a portable system or one that is intended to be
removed periodically—for security reasons or for use away from the
car—a quick-disconnect plug or an adapter that lets the phone take
its power from the car's cigarette lighter may be used.

Control Head

The part of a mobile cellular phone called the *control head* consists
of the receiver and cradle with associated electronics, and is used to
issue commands to the system. It is the part of the system you most
frequently come into contact with since it includes the telephone
handset. The control head may be a one-piece unit, built entirely into
the handset, or may consist of two or more pieces such as a handset
and a small console containing the phone's keypad and loudspeaker.
The control head is connected to the cellular phone's transceiver/
logic unit by a multiwire cable that carries audio and control signals
between the two. It is not required in self-contained units.

The control head issues commands to the rest of the system. It may be built entirely into the handset, or may consist of two or more pieces such as a handset and a small console. (Photo courtesy of Motorola, Inc.)

Transceiver/Logic Unit

The heart of a cellular telephone is its *transceiver/logic unit*. You may think that the handset of a mobile phone is the core of the system, but the real action is in the transceiver/logic unit. A *transceiver* is a combination radio transmitter and receiver (hence its name), and it is the means by which signals travel between your phone and the cell site.

The logic unit is the part of the phone that contains the "smarts." It communicates with the equipment at a cell site to establish a connection, to determine what frequencies will be used for transmitting and receiving, and to coordinate its end of a handoff, when you leave one cell's area of coverage and enter another.

The intelligence built into a mobile phone's transceiver/logic unit is also used to control that phone's power output, just as in portable phones. If the equipment at a cell site senses that it is receiving a strong signal from a phone, it sends a signal telling the transceiver that it can cut back its power. Similarly, if the signal from a cellular phone begins to weaken as the phone is moved farther away from the cell site, the transceiver receives an instruction to increase its output. This ability to vary power output serves two purposes. First, keeping transmitter power to its usable minimum reduces the potential of a phone's signal getting into another cell site on the same frequency

A picture of a car installation. Small transceivers no longer need to be mounted in the car's trunk. (Photo courtesy of JSM Communications Inc., Sheboygan Falls, Wisconsin.)

and creating interference. Second, reduced power output means a longer life for the batteries in a portable or transportable phone — more hours of operation per charge.

In a permanent car installation the transceiver/logic unit is usually mounted in the trunk and connected to the control head and car battery by cables. This location is chosen because it places this important piece of equipment in a safe place and also keeps it from getting in your way. Such a location also places the transceiver close to the antenna, which is usually mounted toward the rear of a vehicle. This proximity allows the length of the cable that connects the transceiver to the antenna (the *feedline*) to be kept short. This is important when working with radio signals at the high frequencies used by cellular equipment, because the shorter the feedline, the better your reception will be. Also, the signal being transmitted will be stronger. With the progressive reduction in the size of cellular equipment, the transceiver may conveniently be mounted in a permanent or semipermanent position in the passenger compartment, if desired. The transceiver/logic unit part of portable cellular phones, or of the transportable phones intended to be operated away from a vehicle as well as in it, is either built into the all-in-one case of the portable or designed as a small unit that can readily be connected to (and disconnected from) a vehicle's antenna and battery. It frequently includes a means for semipermanent mounting near the driver's or user's position.

Antennas

There are several types of antennas for use in a vehicle installation (Figure 8.2). If you plan to have a cellular telephone installed in a vehicle or to have a "hands-free" kit installed in a vehicle for your portable cellular phone, you need to consider what kind of antenna you want. Each type represents a trade-off of ease and cost of installation, performance, and esthetics.

The most visible characteristic of many mobile antennas is the little "pigtail" part way up its length. This served as a "signature" of cellular users before portables came on the scene. This is more than just a decoration. It's called a *phasing coil*, and serves to divide what looks like a single antenna effectively into two antennas, increasing its efficiency.

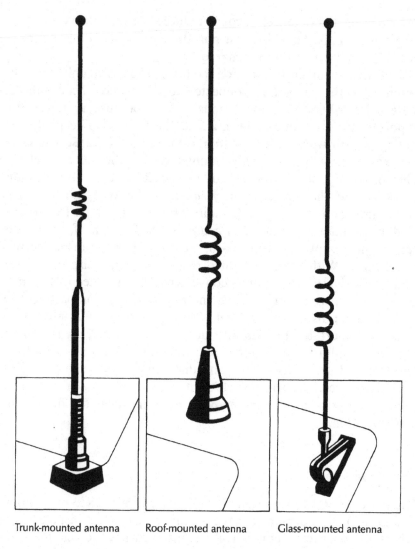

Trunk-mounted antenna   Roof-mounted antenna   Glass-mounted antenna

Figure 8.2   ANTENNAS
Trunk-mounted antenna. Roof-mounted antenna. Glass-mount antenna.

ROOF-MOUNTED ANTENNAS

Ideally, the place for a mobile antenna is on the roof of your car. First, this is the highest point on the vehicle, so the signals emitted from and coming into the antenna are less likely to be obstructed by nearby objects. Second, to function best, an antenna needs a *ground plane*, a surface that actually works as a part of the antenna system, to "launch" radio waves, even though there is no electrical connection between it and the antenna. The metal roof of a car provides a good ground plane.

There are several reasons, however, why most antennas are *not* roof-mounted. The first is because it's inconvenient to lead the antenna cable into the car from an antenna mounted on top. The second is because it creates an obstruction for garages and car washes. The third is because it does not look as nice as other placements, and is used primarily on trucks and commercial vehicles.

TRUNK-MOUNTED ANTENNAS

The next best location for an antenna is on a car's trunk lid, since, after the roof, it offers the next best alternative for a ground plane. It is also less detracting to the vehicle's appearance. Trunk mounting is much easier and more convenient than putting an antenna on the roof. It also shortens the cable to the transceiver/logic unit in the trunk, reducing loss of signal in the cable. In addition it is possible to mount the antenna with a clip that attaches it to the edge of the trunk lid through the space between the lid and the car body. Some ground-plane efficiency is sacrificed this way, but it eliminates the need to drill a hole in the car.

The trunk-mounted antenna is an *elevated-feed* antenna, where the base of the antenna is extended to raise its height above the trunk, and the signal is injected into the antenna at a point above its base. This improves the performance of the trunk-mounted antenna closer to that of the roof-mounted antenna.

GLASS-MOUNTED ANTENNAS

The *glass-mounted antenna* is mounted on the car's rear window, with no physical connection between the antenna on the outside and the cable on the inside. It is not as efficient as the other types of mounts, but it is easier to install, looks better, and requires no hole to be drilled in the car. As mentioned with portable cellular phones, as cellular systems grow and provide better coverage, performance is not as much of an issue as it was several years ago. So the glass mounted antenna has become the most popular antenna type for most private automobiles.

While some owners like to display their glass-mounted antenna at a rakish angle, it should be positioned perpendicular to the ground for optimum performance.

TRANSPORTABLES

Transportables are cellular telephones in which all of the components of a car phone have been assembled into a self-contained unit adaptable for portable use, such as a briefcase, shoulder bag, or rack. Transportables are full-power, 3-watt units, often adapted from their car phone cousins, designed to be used as portables or moved from car to car. They contain a transceiver/logic unit, handset, power supply, and antenna in one unit. Some units intended for in-car use only skip the battery and use only a cigarette lighter adapter for power, and are called *transferables.*

Many transportables convert easily from portable operation to car operation, complete with hands-free and other options previously found on vehicle-mounted phones only. Your cellular dealer can discuss your needs and show you alternative models that are convenient to use in a vehicle, yet quickly adapt to portable use or transfer to another vehicle.

Sometimes, this configuration consists of a bracket with a handle. The bracket holds a rechargeable battery pack, and the phone's transceiver/logic unit and handset can be attached and removed quickly (see Figure 8.3). In portable use, an antenna similar in appearance to "rubber duckie"-type antennas used with two-way portable radios is usually connected in place of the cable that feeds to the vehicle-mounted antenna. If the initial phone installation is done with this dual use in mind, it is a simple matter to switch between in-vehicle and portable use.

If you use your transportable phone both in your car and out of it and plug it into the cigarette lighter receptacle to take advantage of the car's larger battery, find out whether that receptacle is wired into the same circuit as the ignition switch. If it is, power will be available at the receptacle only when the ignition is on. Your phone won't work and its batteries will not charge if the ignition switch is

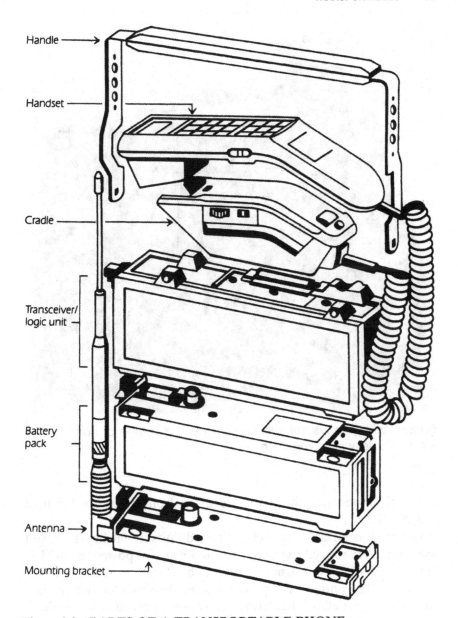

Handle
Handset
Cradle
Transceiver/
logic unit
Battery
pack
Antenna
Mounting bracket

Figure 8.3   PARTS OF A TRANSPORTABLE PHONE

off. If this is the case, don't expect to leave your phone in your garaged car overnight and find it fully charged in the morning. On the other hand, if your car delivers power to the cigarette lighter receptacle with the ignition off, you can drain your battery if you

Transportable phones are intended for use in a vehicle or away from it. (Photo courtesy of Motorola, Inc.)

leave the phone on while you are away from your car for long periods.

Other transportables are self-contained units in hard cases or soft bags, made for rugged use with the full 3 watts of power required in fringe areas where construction or rough treatment in field use are indicated. Bag phones are great for over-the-shoulder use in field applications. An additional advantage of some units is that they look very much like desktop phones, and are perfect for field offices where immediate or temporary service is required.

The major advantages of transportable phones over portables are the full 3 watts of power (which can be automatically reduced when not required to the 600 milliwatt level of portables), rugged construction for professional field use, lower price, and some have longer

With a transportable phone, you can keep in touch with your office from a remote job site. (Photo courtesy of Motorola, Inc.)

battery life than portables. As cellular systems mature and provide better and better service, and portable phones and batteries continue to improve, these advantages are not as important as before, and most people are opting for portables instead. The disadvantages are the size and the weight, typically 2.5–4 pounds.

9

OPTIONS AND ACCESSORIES

Cellular telephones have so many features built into them that it's difficult to think of anything more that could be added to them. Many of these features have been discussed in previous chapters, but we will explore them further here.

BATTERIES

A second battery is perhaps the most-requested accessory of any portable cellphone user. It extends the time between charges while on the road, and ensures that there is always a charged battery at hand if you find that the battery in your phone is dead. Since the standard battery is usually a nickel–cadmium (NiCad) battery, users often will purchase a second battery for more performance rather than just as a spare, and get a nickel-metal hydride (NiMH) or lithium-ion battery. Note that some types of chargers require that the battery is installed in the phone while it is charged. This means that you can't use the phone while it is charging.

CHARGERS AND ELIMINATORS

Battery chargers may be *trickle chargers* or *rapid chargers*. Rapid chargers are more expensive and charge the battery faster. *Travel*

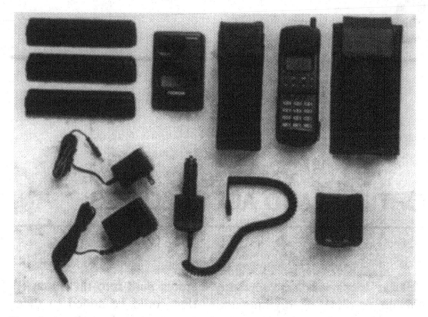

Popular accessories include a case, extra battery, battery charger, and battery eliminator for the car. (Photo courtesy of Nokia.)

chargers are small trickle chargers made to take up less room on trips. Some premium chargers may charge several batteries at once, and serve as a convenient cradle for the phone at home or office, charging one battery while it is installed in the phone.

An important issue for portable users is battery life, so a battery eliminator is an attractive accessory. The battery eliminator plugs into the cigarette lighter of a vehicle and substitutes for the battery. Most also charge the phone's battery at the same time.

HANDS-FREE OPERATION

When you were learning to drive you were told to keep both hands on the wheel! When you started using a cellular phone, you discovered you had to use one of those hands to hold the phone. The handsets of mobile phones are not designed to be "crooked" between

A hands-free mobile telephone. (Photo courtesy of Clearvox Communications.)

your neck and shoulder, since holding your head at that awkward angle would make it extremely difficult to concentrate on the road.

Most manufacturers of mobile and portable phones offer hands-free kits that make it unnecessary for you to hold the handset while carrying on a phone conversation.

The hands-free arrangement consists of a speaker, which may be built into a mobile phone's control head, and a small microphone, which clips onto your sun visor or other convenient place to pick up your end of the conversation (see Figure 9.1). Some factory-installed systems include installation of the microphone and some memory-dialing capability in the steering wheel, sun visor, or other convenient location. Some systems allow you to switch between hands-free and handset operation during a call if the car or connection is noisy, or if the person on the other end asks you not to use the "speakerphone."

There are other features that permit you to operate a cellular phone with minimal interference with driving. *Voice Dialing* may be available from the carrier or built into the phone, allowing you to dial numbers by voice, or by calling out a name whose associated number you have previously programmed into the system. Voice-messaging systems of cellular carriers often have procedures you can use to check your messages using abbreviated number dialing and minimal keystrokes.

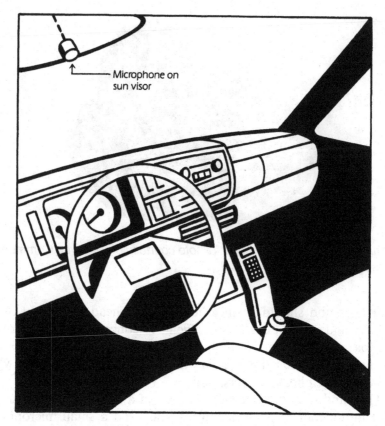

Microphone on
sun visor

Figure 9.1 ONE TYPE OF HANDS-FREE OPERATION

PAGERS

Paging service is not directly related to cellular service, but is an important accessory for many users. With a pager, cellular users can always be in touch, while managing their incoming cellular calls. Such calls may occur at inconvenient times and places, incur expensive airtime charges, and require cellular users to give out their cellular phone number to people to whom they do not always want to be available. Paging service can be local, regional, or national, as necessary, and when paged the user can determine the appropriate time and place to return the call, perhaps even waiting until a less expensive landline phone can be reached. Paging service can often be arranged through your cellular salesperson.

... Plus I want to buy 200 shares of CTM Systems and sell all of the Acme Chemical, then convert the ...

CASES

A vinyl or leather case is an attractive and practical accessory for a portable or transportable phone. These are available from the manufacturer as well as many custom case manufacturers; some are included with the phone. As the price of phones has come down, many users don't see as much of a need to protect the phone. Some cases may cover the keypad or otherwise make the use of the phone less convenient. Consider cost and convenience, along with appearance, when selecting a case.

DATA APPLICATIONS

Telephones are not the only devices that have recently gone portable. Computers have also become portable, with some offering all the power of desktop systems in packages little larger than a loose-leaf notebook.

One use to which these *laptop* computers are frequently put is communicating with the Internet, or with other, larger computers — either to send information (perhaps a day's orders taken by a salesperson working in the field) or to receive it (for example, to check on the status of an order already being filled).

People on the move find that cellular allows them to use data communications where traditional telephone service is not available. (Photo courtesy of Nokia.)

All this computer-to-computer communication takes place by telephone. Digital signals output by your computer are converted into audio tones by a device called a *modem* and sent to the distant computer over standard telephone lines, where they are converted back into digital form. Information from the remote computer is transmitted back to you the same way. Many personal computers (PCs) and laptops come equipped with modems.

Data applications are gaining in importance among users of wireless services, especially with increased interest in information services, e-mail, and the Internet for both business and personal use. Many people find that wireless access to data applications is as important as the freedom and productivity that cellular provides to their voice communications.

If your PC or laptop has an existing modem, to use it on a cellular system at the very least a device is needed between the modem and the cellular phone. This device makes the cellular phone appear to the modem as an ordinary phone with dial tone, which cellular doesn't have, as well as an adapter for the standard RJ-11 jacks which most modems have. These devices are appropriately called *RJ-11 Dial Tone Interfaces.*

Special packet data service for cellular, called CDPD (Cellular Digital Packet Data), is also available in most areas of the country, which provides low-cost wireless access for uses like short e-mail messages and special commercial applications. Other forms of wireless data access besides cellular are also available.

A CELLULAR DATA COMMUNICATIONS SESSION

Your data interface or cellular modem should have a plug made to fit a jack on your cellular phone. Once this is connected, you should be able to dial a data call and press SND just as easily as you place a voice call.

There are several problems that can arise in cellular data communications, and cellular modems are designed to take them into account. The biggest obstacle to reliable data communications is signal dropouts (see Chapter 10). These sometimes occur even on conventional telephone lines and may also be encountered in cellular systems. In data communications, an interruption of a couple of milliseconds — just a few thousandths of a second — can be disruptive. So can bursts of noise, which the computer may attempt to treat as data even though they are nothing of the sort. Even the infinitesimal, and usually unnoticeable, interruption that may take place as a call is handed off from one cell to another can wreak havoc with data communicated by modem.

The errors that may be induced by signal fading, dropouts, or noise cannot be eliminated, but they can be corrected. There are two ways in which this can be done.

The first method uses an error-correction scheme built into the modem. As the modem sends out data, it also periodically transmits a number that it has calculated based on the values of the numerical representation of that data. The modem at the receiving end performs the same type of calculation on the data it receives. If the figure it arrives at disagrees with the one that has been sent with the data, it knows that an error has crept in, and requests that the information be transmitted again. The chances are that the second time it will come through correctly.

This error correction by modem is particularly useful when information is being sent from the keyboard of one computer to another. It can't fix typing errors, but it does ensure that what gets typed is received correctly at the other end.

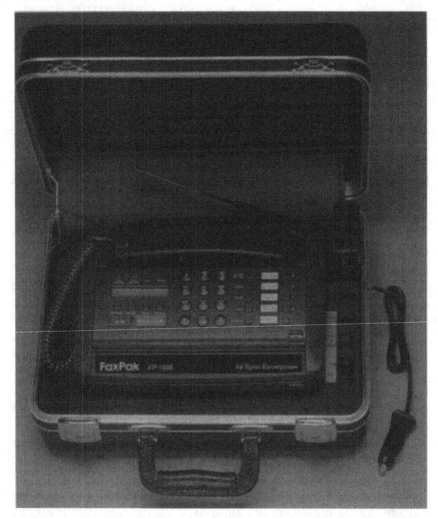

With a portable fax machine you can send and receive documents via your cellular phone. (Photo courtesy of Ryno Enterprises, Long Beach, California.)

Information can also be sent from computer to computer directly, without entering it from the keyboard. Files (collections of data) can be created and stored on floppy disks and then transmitted from one computer to another all at once, which is much faster than they could be transmitted if they all had to be typed in character by character at the time of the transmission. Sometimes this is the only way certain information can be sent; at other times it simply provides the means to cut down on connect time and keep airtime charges to a minimum.

Ordinary (deskbound) modems are quick to hang up when they no longer hear the signal from the modem at the other end of the line. When one modem hangs up, so does the other. With dropouts and handoffs being a fact of cellular life, this could present a problem. Consequently, modems intended for cellular phone use are more tolerant of signal loss than regular modems, and can hold the connection longer when the transmission is interrupted.

Your carrier may have a specific prefix or dialing procedure to accept calls from cellular modems, to enable error correction to work most effectively. It will direct your call through a cellular modem at the mobile switch center (MSC), allowing you to utilize the benefits of the cellular data-error-correction protocol during the wireless portion of the transmission. Then it will be converted to standard modem protocol. This eliminates the need for a special cellular modem and incoming line for cellular calls at the computer or service bureau you are linking up with.

Cellular modems have advanced in speed capability as fast as their landline equivalents. Like landline modems, cellular modems can work at speeds in excess of 14,400 bits per second (bps) over regular phone lines. However, since cellular introduces several additional variables that can increase error rates over landlines, 1200–9600 bps is the norm for actual throughput.

Needless to say, it would be extremely unwise to attempt to use a computer and a cellular phone together while driving (see Chapter 11, "Safety and Security").

FAX APPLICATIONS

RJ-11 dial tone interfaces can also be used to interface your cellular phone to a fax machine. Portable fax machines are available that you can use anywhere you use a cellular phone, and may have a built-in cellular modem as well as battery power.

Some voice-messaging services (see voice messaging as a service option elsewhere) permit you to receive incoming faxes also. When you check for messages, you can have the faxes print out on your portable fax, or redirect the faxes to your next destination. This feature allows a lot of flexibility to those who use their automobile as a "mobile office."

SHORT-MESSAGE SERVICE

Many cellular services now have a service available that enables you to receive messages on the display of your phone. You can be notified of voice messages you have received, get stock quotes and weather information, or use it like a pager. You may even be able to send short messages to other users using the Internet or a service you can call. Your salesperson can show you a phone that includes a display and short-message service (SMS) capability, and explain the SMS services available from the carrier he or she represents in your area.

10

DEALING WITH OPERATIONAL DIFFICULTIES

Cellular telephones are rugged and reliable. They have to be, since their very nature dictates that they will be banged around and subjected to extremes of heat and cold in briefcases, automobiles, and other places. They rarely fail. Most of any difficulties with your cellphone will arise from the conditions under which it is used and, possibly, from your initial unfamiliarity with the equipment and with cellular systems in general. This chapter will help you to understand some of the difficulties you may encounter in using your cellular phone, and, when possible, will show you what you can do about them.

BAD CONNECTIONS

As with any phone system, once in a while you will get a bad connection with your cellular phone. This is particularly true in fringe areas — near the edge of a company's area of service or where the terrain interferes with the system's radio signals. Often it is possible to continue a conversation under poor conditions, but sometimes things get so bad that you are automatically disconnected.

Disconnects occur because of signal dropouts — either your phone loses the signal from the cell site or the cell loses the signal from your phone. These two things do not always happen simultaneously;

sometimes you will be able to hear the other party, but they won't be able to hear you, or vice versa. Sometimes there is a way to tell when a disconnect is going to happen, allowing you to exit gracefully from a call before you are cut off.

All phone systems—landline and cellular—provide some sort of audio feedback. That is, when you speak, you can hear a portion of the electrical signal (your end of the conversation) in the earpiece of your phone. This phenomenon is sometimes called *side tone.* You're usually not conscious of this little voice whispering in your ear, but you always hear it, and it reassures you that your phone is operating properly.

When you don't hear this feedback signal, you know you're in trouble. Those times when you pick up a phone and its dead, that "deadness" you hear is due to the absence of the feedback signal. Try this experiment on your home phone. Pick up the receiver, put it to your ear, and blow into the mouthpiece. You'll hear yourself blowing in your ear, as it were. Now unplug the phone and do it again. You'll hear the difference immediately—no feedback.

Sometimes when you're speaking on a cellular phone you'll notice a momentary loss of feedback as you talk. This indicates either that the cell site lost your signal for a moment or that you lost its signal.

Cellular phones are designed to take these momentary dropouts into account and will not disconnect unless a dropout exceeds a certain period. However, if you experience a number of dropouts one after the other, be prepared to be disconnected. Multiple dropouts are a sign that you're in an area of poor signal reception or transmission and that there may be more, one of which may be long enough to cause the system to disconnect you, thinking you're no longer there. Should you experience these dropouts, tell the party at the other end of the line that you may be disconnected from them abruptly and that you'll call them back if this happens.

SIGNAL DROPOUTS AND DEAD SPOTS

Signal dropouts are annoying. They can ruin conversations by causing you to lose words, by having to repeat words, and by causing disconnection. *Dead spots* are regions where your cellular phone doesn't work at all, and *dropouts* occur when you run into a series of dead spots. There isn't much you can do about these when you encounter them, but there are steps you can take to avoid them.

What causes dropouts and dead spots? The answer lies largely in the behavior of radio waves at the frequencies used by cellular phones. At these frequencies, almost a billion cycles per second, radio waves start to act a bit like light waves. They travel in straight lines, can be weakened by water in the air (which is why UHF TV signals aren't as strong on rainy days), and are easily reflected by man-made and natural objects (the same principle that makes radar work).

For reasons that are not entirely clear, some areas are just naturally dead to radio waves. This may be due to the terrain or to disturbances in the signal path caused by objects many miles away; whatever the cause, some areas are just not good for reception or transmission at certain frequencies.

Dead spots are frequently encountered in mountainous or hilly terrain and sometimes in heavily wooded areas. But they are not restricted to rural areas (see Figure 10.1). There are many dead spots in cities. These have two causes. The first is simply the blockage of signals by buildings between your phone and the cell site. Although the site antennas are located as strategically as possible for optimum coverage, there is bound to be something in the way somewhere.

Dead spots also occur because of *multipath reception* (or just *multipath*), which happens when two or more signals interfere with one another. The signals may be different or they may have the same source (see Figure 10.2).

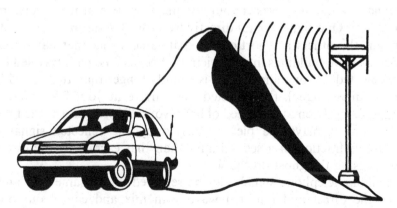

Figure 10.1 DEAD SPOTS CAN OCCUR IN MOUNTAINOUS TERRAIN

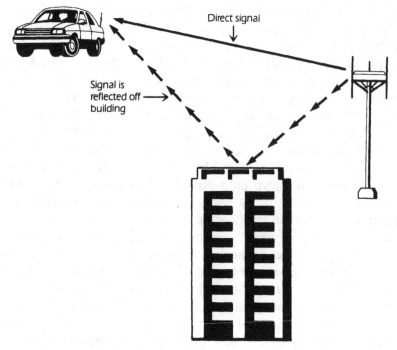

Figure 10.2   MULTIPATH RECEPTION
Multipath reflected waves can mix with each other (and with direct signals), resulting in dead spots or badly distorted reception.

Direct signal

Signal is reflected off → building

One example of multipath is the "ghosts" on your TV screen. The original TV signal arrives at your antenna from two or more different directions. One signal comes directly from the TV station's transmitter, but the others are portions of that same signal that have been reflected from buildings in your vicinity. Because the path traveled by the reflected TV or radio signals is slightly longer than that traveled by the direct signal, the reflected ones arrive at your TV antenna slightly later. Even at the speed of light, your TV set discerns the time delay and displays the picture carried by the reflected signal a minuscule fraction of a second later than that carried by the direct one, causing the ghost on the screen.

Cellular telephone signals can be reflected in the same way and, since the direct and reflected waves can mix and cancel out one another, communications can thus be seriously affected (see Figure 10.3). When the peak of one radio wave mixes with the trough of

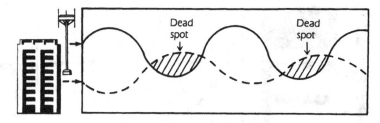

Figure 10.3   DEAD SPOTS FROM MIXED SIGNALS
Where the top of a direct signal mixes with the bottom of a reflected signal, a dead spot occurs.

another, the result is effectively zero — the combination results in no signal at all, or a badly distorted one. If you are at the spot where this happens, you will be in a dead spot. Depending on conditions and which frequency your conversation is currently randomly assigned to, dead spots can be small or huge.

Dropouts happen when you pass through a number of dead spots, one right after the other. The conditions that give rise to one dead spot frequently give rise to many.

What can you do about dead spots or dropout areas? Not much. If you routinely travel the same route, try changing your path by a block or so, or try not to use your phone when you're in a trouble spot. You may find, especially in the case of multipath, that simply moving a foot or two will restore signals both ways (that's why you can make TV ghosts shift or even disappear by adjusting an indoor antenna slightly). If this doesn't work, you'll just have to wait until you're in a better place.

Dead spots are frequently responsible for causing a phone's NO SERVICE indicator to light. This light comes on not because there's no signal in the city you're in but because, since there's no signal in its immediate area, it *thinks* it's in a no-service area. And, in a sense, it's right.

There is, however, a bright side to the multipath phenomenon. Sometimes, even if the direct signal is blocked, you may be able to use your phone via a reflected one (see Figure 10.4). There is no rule to guide you in this, but you'll often find that you can make and receive calls from the most unlikely places.

Cellular carriers continue to find ways to provide signal in some unlikely places. Tunnels and subways have traditionally been areas

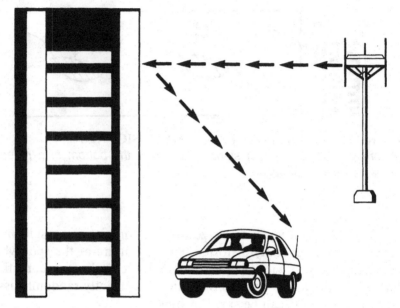

Figure 10.4   REFLECTED SIGNALS
Even if a direct signal is blocked, your phone may be able to use a reflected signal.

where no one would expect to find cellular service available. Yet, advances in technology have permitted carriers to find ways to provide coverage in these areas.

One additional problem may cause you to encounter areas of poor signal. If you are heading away from a cell site and the system gets ready to hand off your call to the next, the next cell may be fully occupied on all channels with calls in progress. The signal from the current cell may continue to deteriorate beyond the point where it would normally hand off the call to the next, causing all the symptoms we've described earlier in this section. This condition would not occur on a regular basis, though the others might.

OTHER INTERMITTENT EFFECTS

When placing a call from an area of marginal service, your phone's IN USE indicator may come on and then, after a few seconds, be replaced by the NO SERVICE indicator. This happens because, when the phone makes contact with the cell site, it (or the cell site) decides that the signals are not good enough to maintain a connection and disconnects you. There's nothing to do about this but to move elsewhere.

Similarly, you may find that, even though you can carry on conversations when passing through certain areas, you can't place calls from them. This is because the thresholds of the cellular system are set to stop a call being initiated when the chances are that it won't be successful, but to allow an in-progress call to continue as long as possible.

Dead spots come in all sizes and intensities of deadness. "Picket fencing" or "flutter" happens when you drive quickly through alternate zones of strong and weak signals, causing the signal strength at your antenna (or at the cell site) to fluctuate up and down. If you slow down, the rate of flutter may also slow down. You can usually carry on a conversation under these conditions, suffering nothing other than a little annoyance until it passes. With digital service, you may experience muting instead of the sounds described here for regular analog cellular service.

Even though the FM radio signals used by cellular phones are immune to most interference, you may still occasionally experience some static, particularly in some metropolitan areas. This is often due to interference from nearby powerful transmitters, such as those used by broadcast stations (New York City's infamous Intermod Alley is located directly beneath the Empire State Building, which bristles with broadcast and other commercial service antennas). You can't do anything about this except to get out of their way, but it's reassuring to know that the difficulty is not being caused by your phone.

Static may also interfere with your conversation when you are near the limit of a region's coverage with no adjacent cell site to be handed off to. As the signal level drops, outside noise, normally covered up by the cell's signal, may sneak in. Should you encounter this, tell the person with whom you are talking that you may be suddenly disconnected if the cell site signal drops so low as to be useless.

Indoors is one place you might not expect to have difficulty in using your cellular phone. Yet, in certain buildings—even if they're right under a cell-site antenna—you may find your NO SERVICE indicator lit, or find yourself disconnected when you talk while carrying your portable phone from place to place.

This is because the steel used in high-rise buildings for beams and girders, and for the reinforcing rods and mesh used to strengthen concrete, works as a radio-wave shield. Notice how the performance of broadcast-band AM and FM radios falls off in certain parts of office and apartment buildings. Reception may be good near the outside walls, especially by windows, but it becomes more difficult the deeper into the building you get. The effect is the same on cellular phones. It is these same materials, by the way, that reflect radio signals and both help reception in some areas and sometimes cause multipath interference.

If you must use your cellular phone in a metal-frame building (wood-frame construction won't faze it), try it near a window, preferably one facing in the direction of the cell site you're using. Or try positioning your phone near or under an air-conditioning duct. The duct acts as a pipe for high-frequency radio waves; just moving your phone six inches or so in one direction or another may help you to establish a connection or clear up a noisy one.

MECHANICAL OR ELECTRONIC FAILURE

A time may come when your cellular phone refuses to work. Before you blame the manufacturer or dealer who sold it to you, try to find the source of the problem yourself. It may be — and usually is — something that's completely unrelated to the phone proper, and you can often fix it yourself with a minimum of fuss.

Power Problems

When you turn your cellular phone on and nothing happens — no lights, no reassuring beeps as the unit checks itself out — it's probably a lack of power. This usually happens because the battery is discharged or is improperly installed and making a poor connection to the phone. If your phone is powered from the cigarette lighter receptacle of your car, make sure the ignition switch is turned on. In many cars, power to the lighter is controlled by this switch, and if the ignition is off, the phone won't work and its battery won't charge.

You may also have accidentally locked the phone, or forgotten that you locked it. If you don't often use this feature, check the manual for the procedure to unlock it.

No Service

If you find you're getting a NO SERVICE indication when you know you're in a good signal area, or if the operation of your phone seems to be deteriorating, the problem may be your antenna or reception. If you are using an external vehicle antenna, make sure you haven't forgotten to replace it after removing it for a car wash or other hazard. There are occasions when areas with normally good reception may have a temporary cell-site outage or a transient or indoor signal interference problem.

If you can't make a call, you may have entered the number incorrectly, or are experiencing a temporary system blockage or outage. To guard against fraud some systems require a personal identification number (PIN) code to make a call, and you may have forgotten to enter it or entered it incorrectly.

A-B Switch

Although the A-B switches of most phones can be programmed for a priority mode—that is, they will always look first for an A or B carrier, depending on how they are programmed, and then for the other only if there is no service from the priority carrier—it is sometimes better to lock your phone into the service you normally use at home, with no secondary choice. Then, if you roam to another city, switch it if necessary from A to B or vice versa if you have a carrier preference when roaming. This will ensure that, if you enter a spotty region within your home area, your phone will not accidentally lock onto the service you don't subscribe to, which would get you no service at all, or a recording telling you to call that carrier's business office to get service.

If your phone is locked into a single A or B mode when no service is provided on that block at the moment, you will get a NO SERVICE indication. Try switching to the alternate service, just to make sure there is nothing wrong with the phone. You may get a ROAM indicator and not be able to place calls, but you will know your phone is working correctly.

SERVICE AND REPAIR

Even the most reliable equipment may suffer a component failure. The conditions under which some cellular phones are used, and the ease with which they can be left in places where they may suffer water or heat damage, may eventually cause it to malfunction. At some time your phone may start to act up or refuse to work at all as a result of an internal electronic failure. Unless you have the training and lots of expensive test and alignment equipment to back you up, there is little you can do to correct such a failure.

If, after making sure all the phone's connections are secure and checking for other obvious sources of trouble, you decide that whatever is ailing your phone is nothing you can take care of yourself, return the unit to your dealer or to another source of reliable service.

A well-equipped service shop will check out your phone, isolate the problem, and correct it in a matter of a day or two. Most cellular phones are designed on a modular basis, with various parts of their circuitry situated on individually replaceable circuit boards. Once a problem has been located, a defective module can be removed and a good one installed. Often, the phone's warranty will cover the cost of repair.

In the unlikely event that your phone has to go into the shop, your carrier may provide you with a "loaner" to make sure you have a phone to use until yours is repaired. It is in their interest to keep you in service as much as possible.

11

SAFETY AND SECURITY

This chapter concerns safety and security—*safety* in the sense of maintaining your usual standard of driving while using a cellular phone, how cellular improves safety, and in safeguarding your phone equipment; and *security* in the sense of guarding the privacy of your conversations.

SAFETY

Anything that distracts you from driving can be a hazard. A cellular phone, unless it's used with care, can be just such a distraction. Using a phone at the same time you're trying to concentrate on the road and traffic conditions can endanger both you and other drivers.

For this reason, try not to dial a number while you're moving. Dialing takes a lot more attention than you think; your mind, eyes, and hands are diverted from controlling your vehicle. The designers of phones have included features to make the equipment safer; take full advantage of what they have provided.

One thing that makes cellular phones smart is their memories. Some phones can store ninety-nine or more numbers. Recalling these numbers requires only the pressing of a couple of keys and then pressing the SND key to set the dialing process in motion. Even so, your attention can waver during the few seconds it takes.

If you must make a keypad entry while in motion, use the handset in its cradle, if you have one. This position, in the case of a permanent or semipermanent installation, is more or less in your line of sight with the road. Leaving the handset in place at least until your call goes through ensures that you have two hands free for as long as possible.

Some of the phones and add-ons recommended in Chapter 9, "Options and Accessories," make driving while using the phone less of a risk to yourself and others. These safety-oriented devices include phones that place the keypad directly in your line of sight — on the dashboard, for example — and make it unnecessary to pick up the handset until a connection is established. Speech recognition for dialing also is available, as is hands-free operation.

Be especially careful if you're using a transportable phone that's on the seat next to you. Looking over at its handset to verify that you're pressing the right keys takes your eyes from the road completely; you could forget for a moment that you're guiding a couple of tons of iron and steel at close to a mile a minute. Unless you're parked, dial these phones with the handset in front of you, not while it's in the passenger seat. You've almost certainly encountered, at some time in your driving career, drivers who were so completely engaged in conversation with the passenger sitting next to them that they were oblivious to everything else on the road. Bear in mind that this could be *you* when you're using your phone.

Without being aware of it, you may start paying more attention to what's being said or to what you're going to say than to the traffic or road conditions you're in the midst of. It's bad enough to miss your turnoff because your mind was on something else; it will be worse if you turn *into* something you didn't notice was there.

Don't let your mind or attention wander. Make safe driving your priority. If things get busy on the road, hang up and continue your call later.

Protecting Your Phone

A cellular phone is a valuable piece of equipment. The antenna on a car-installed mobile phone or car kit for a portable can signal thieves that there is something inside worth a lot of money. It's simple to prevent them (or anyone else) from using the phone once they get their hands on it — just notify your cellular service provider — but why give them the opportunity to steal it in the first place?

Recently, the most common thefts have been breaking the passenger side window to take a portable cellular phone from the front passenger seat. If you use a portable, take it with you when you leave the car, or at least disconnect it and stow it out of sight in the trunk or elsewhere. If you park anywhere for a while, for example, at an airport, and have a removable car antenna on your vehicle, take it off and store it out of sight. This, at least, will not draw so much attention to your car as a potential target for theft.

To discourage unauthorized use, most cellular phones come with a lock feature. By pressing the LOCK key, or other special key sequence, the phone is set to accept incoming calls, but will not permit dialing out. To unlock it, you must enter a code known only to you. Some phones have two locking systems, one that is stored permanently in the phone's numeric assignment module (NAM), and one that can be changed at any time by someone who knows how (you generally need to know at least the NAM unlock code to access the other).

There are a number of automotive burglar alarms on the market, and, if you store or use your car in a high-risk area with the phone in it, you probably should have one installed. Sometimes the sheer racket these devices make when tripped is enough to scare would-be thieves away. Of course, the first level of security is always locking your car.

The popularity of portable phones has brought a new problem — losing the phone. One security feature of car phones is that the phone is attached to the car, preventing absent-minded users from losing it easily! Portables are too easy to leave at an airline counter or in a rental car. Business users, who need the phone handy all the time, often don't keep it in a briefcase or purse. The only solution to this is to develop a habit of returning the phone to a secure place immediately after use, where it can be easily retrieved for an incoming call. A purse is usually a good bet for a woman. Many male business users are wearing it on a belt like a pager, or keeping it in their briefcase.

SECURITY

Concern over the security of telephone conversation as it applies to cellular phones is probably overstated. Certainly, when a telephone conversation, which is protected by privacy laws, goes over the air, as is the case with a cellular phone link, there is reason to be concerned. However, there is probably much less cause for concern than you may have been led to believe.

Early in this book we noted that some parts of the radio spectrum used by cellular phones coincide with the upper reaches of the UHF TV band. Some high-end radio scanning receivers may also cover these frequencies, but their sale in the United States has now become illegal. The question is, how easy is it to eavesdrop on a cellular phone conversation—putting aside for the moment the legality of the matter—with this equipment?

The answer is, not very. The way cellular phones operate makes their signals difficult to locate and even more difficult to track. To eavesdrop on a conversation, you have to know two things: *when* it is going to take place, and *where*. And, while eavesdroppers may have some idea when the words they are waiting to hear may be uttered, they have no idea where, among the 832 channels assigned to cellular telephony, those words are going to show up.

It's not like tapping into an ordinary telephone line and then sitting back and waiting or listening to the playback of a tape recorder. You must be at the right place at the right time, which is virtually impossible, given the way cellular phones work.

The frequency pair, or two channels, on which a conversation will begin is determined randomly and automatically by a cellular phone system's equipment according to the conditions that prevail at that instant. The location of the cellular phone user determines which cell site (of many) will be used, and each cell site has assigned to it a set of frequencies that differ from those used by adjoining cells. Which frequencies within a cell will be chosen for a particular conversation (or part of a conversation) depends on the ones that are free when the call is made. Further, if a cell site has been split (see Chapter 2), the chances are that the new cells are served by directional antennas, that is, they concentrate their signals in a particular direction. A would-be eavesdropper on the wrong side of the antenna has little hope for success. In addition to the initial problem of finding the correct frequency pair, cellular telephony adds the complication of frequency changes when a handoff is performed. Highly secure government and industrial radio communications use a similar technique (called *diversity transmission and reception*) to scatter a confidential conversation all over the radio frequency spectrum.

As the cellular phone user moves out of one cell and into another—usually only a matter of a few miles, no more than 10 minutes in a car—the responsibilities for the radio link are transferred to that new cell site. And, since adjacent cell sites use different sets of frequencies to avoid interference with one another, the frequencies the conversation is transferred to will differ from those under which it was initiated. Again, frequency selection is done automatically and randomly, and there is no telling where the conversation will show up.

In summary, given that people could obtain an illegal radio to receive cellular frequencies, it is possible for them to intercept cellular telephone conversations. But the possibility that they could hear the conversation of any *particular* person is minimal, and all they would hear is random snatches of conversations from random users.

Security Devices

Despite the extreme unlikelihood of anyone's coming across—and being able to track—your cellular telephone conversations, you may feel you need some measures to prevent your privacy from being compromised.

The best way to keep secrets from leaking is not to discuss them. When you are discussing matters of a sensitive nature on your cellular phone or when you think they may be mentioned, remind those at the other end that they are participating in a cellular phone conversation, a portion of which is going out over the air. Reminding them that their conversation potentially is open to public ears can prevent indiscretions.

If you *must* talk about private matters, there are devices to ensure that they stay that way. The first is a clamp-on unit that you attach to your phone's handset. This small, lightweight unit is powered by a self-contained battery and can be used with almost any phone. It works on the principle of *audio inversion*, intercepting the sounds that form words and changing their characteristics so they are unintelligible without a reinverting device. The characteristics of the audio inversion process can be modified by changing the settings on a small switch in the voice scrambler; there are usually tens of thousands of combinations. Only the same combination set on an identical unit attached to the phone at the other end of the conversation will produce an accurate reproduction of the original speech.

This type of scrambler is, by current standards, a relatively unsophisticated device. Still, given the already built-in safeguards against eavesdropping that cellular phones provide, it should afford you all the extra protection you feel you require. There are, however, more elaborate protection devices available. These scramblers use digital techniques and complex encryption schemes to provide the utmost in privacy. Some can handle both speech and data.

Devices of this sort are usually owned by the parties using them. They must prearrange their phone call, have the devices ready at the time of the call, and of course no one else can participate in the conversation. The information they transmit is scrambled over the entire path between one phone and the other, including the landline and radio portions. Some phone systems, however, may offer a service that requires only the cellular unit to have scrambling/descrambling equipment. The information transmitted, be it voice or computer data, is sent in encrypted form over only the portion of the phone link that uses radio. Once it arrives safely at the mobile switching center (MSC), it is decrypted by on-site equipment and completes its journey over ordinary landline in unencrypted form. No conversion device is required at the receiving phone, which means that a sensitive call can be made over a cellular link to anywhere.

The process is two-way; what comes from the office- or home-bound phone is encrypted at the MSC and decrypted by the equipment associated with the cellular one. The cellular service you subscribe to can tell you whether it offers this or a similar protection scheme.

Finally, digital cellular and personal communication service (PCS), which we discuss in the next chapter, accomplish the scrambling function automatically. As long as you can arrange that the other persons on the conversation are using digital phones in digital mode, your conversation should be secure.

Cellular Fraud

The stealing of cellular service is illegal. Some criminals may use a fraudulent name to obtain service, or steal identification to pose as someone else. This is called *subscription fraud*. The main type of fraud is *cloning fraud*, and a more sophisticated type of cloning fraud called *roaming fraud*. These are discussed in the chapter on roaming.

12

INTO THE FUTURE

The technology that permits us to put telephones in our cars and in our pockets continues to evolve and provide us with new and improved services. Because there is no need for utility poles, conduit, or miles of expensive cable, cellular telephones can be located, either permanently or temporarily, wherever they are needed. This versatility gives cellular the crucial economic edge that ensures the rapid growth of this new industry. In this chapter we will look into the future of wireless telecommunications and its applications.

DIGITAL CELLULAR

New technology permits the radio portion of cellular to become digital. From the carrier's perspective, digital technology allows more channels than analog, and thus greater capacity in the cellular system. From the user's perspective, digital transmission will provide several advantages. First, digital encrypts the conversation automatically, providing even greater privacy of communication. But more importantly, digital provides clearer audio and more consistent radio communication. Besides improving the audio fidelity of communication, the background noise level will be reduced, and static, interference, and other competing noises that occasionally occur in the transmission are virtually eliminated, making the connection much clearer. Finally, and perhaps most importantly, batteries last much longer on a charge using digital cellular. This is a very important advantage to people who use their cellular phone frequently.

Digital cellular is probably available in your area. In order to make the transition, cellular phones that are capable of both standard analog, as well as the new digital transmission, are usually used. These are commonly called, "dual-mode" cellular phones, and are available in all types: portable, transportable, and mobile. While analog phones are not going to become obsolete soon, investigate these phones when you consider buying. In order to accelerate the transition to digital, carriers may have special offers on both digital phones and service that will minimize the cost of having the latest technology available to you.

There are two competing standards for digital cellular. Time Division Multiple Access (TDMA) causes different conversations to take up different time slots on the same channel. A competing technology, Code Division Multiple Access, or CDMA, "tags" digital pieces of the conversation with a coded address and sends them over a broad range of frequencies, to be assembled at the other end by a receiver that uniquely is identified by the code, which changes during each transmission. This technology allows even more economical use of the channels, and may provide even more capacity.

Both CDMA and TDMA provide the same advantages to users. Both standards have been approved, but they are not compatible with each other. Therefore, you may not be able to use digital while roaming to certain cities if the carriers do not adopt the digital standard used on your phone in your home city. But in that case, your phone will revert to analog operation, and you can still use it.

The premium charged for digital phones is not great, and many carriers charge the same or similar rates for digital service as for analog. The phones have virtually the same appearance and work the same way as regular cellular phones. Consider this option when you decide on your phone and carrier.

PERSONAL COMMUNICATIONS SERVICE

The latest in wireless telecommunications service is personal communications service (PCS). PCS offers cellular-like service, using digital technology. At 18–1900 MHz, it operates at a much higher frequency than cellular. Large cells are impractical at this frequency, but PCS uses many small cell sites, or microcells, operating at low power.

Since the service requires many more cells to cover a given area, it is more difficult and expensive with PCS to completely cover a given geographic area than with cellular. On the other hand, once an area is covered, there should be fewer dead spots, and more capacity than with more widely spaced cellular transmitters.

PCS got its start when the Federal Communications Commission (FCC) authorized frequencies for additional wireless services, and auctioned them to make money for the government, as well as to avoid a lengthy process of evaluating who might be the most qualified applicant to provide this service. There are up to six PCS carriers in each area, providing a lot of competition for cellular. To make things even more confusing, some cellular carriers refer to their digital cellular service as PCS, as a competitive weapon against PCS.

Like digital cellular, there are several competing standards for this service. In addition to the CDMA and TDMA technologies, which cellular has, there are other standards, including DCS 1900. This standard is an up-banded, or higher-frequency version of the European digital cellular standard, Global System for Mobile (GSM), which has become a worldwide standard for cellular outside the United States and Japan, which each have developed their own. Global System for Mobile is derived from the French term for the European standards group that developed it, *Groupe Speciale Mobile*.

Carriers may provide a "dual mode" phone for PCS, like digital cellular usually provides, which permits users to switch from one standard to another if they roam to another city. PCS may offer additional features and services (although cellular carriers may offer them also), including a pagerlike service that shows alphanumeric messages on the display. This is called short message service, or SMS (also available from some cellular carriers; see Chapter 9, "Options and Accessories"). Some carriers may charge lower prices, provide additional services, or remove the requirement for a long-term contract, in order to make PCS aggressively competitive and to gain market share against the large, entrenched cellular carriers in the same area. The major variables for considering PCS along with cellular as your wireless telecommunications service (the more general term encompassing cellular, PCS, and other wireless voice telephony services) include price, coverage, and roaming, if the latter

is important to you. PCS should generally provide the same benefits and value as cellular. Additional services, such as Enhanced Specialized Mobile Radio (ESMR), may also be available as an alternative to cellular in your area.

TECHNOLOGY

Cellular telephones, like all things electronic, are getting smaller. Although handsets are already as small as they really need to be (there is a limit to how small a keypad can be made and still be used by normal-sized fingers), phones that are "wearable," like a watch or in a small holster like a pager, are available. Thus, phones will become even smaller, but more importantly, will be more adjustable to the lifestyles of the increasingly diverse cellular user community.

The most significant hardware advances in phones are being made in battery technology, permitting longer battery life between charges for a given battery size. Displays are becoming larger and more useful in providing more services, as well as in helping the user operate the many functions without the need for a user manual.

For the cellular system, in addition to the advances in digital technology, much work is being done to make cell-site antennas more efficient and less obtrusive to the environment. Manufacturers are

This palm tree is actually a fully operational cell-site antenna that is an attractive addition to the surrounding scenery. (Photo courtesy of Valmont/Microflect.)

able to disguise antennas as trees or buildings, without compromising the performance of the system. And new "smart" antennas can dynamically change their radiating pattern to make coverage and capacity more adaptable to changing conditions.

In the near future, cellular systems will be able to report your location when you call 911. This location feature will also enable additional applications, like finding the nearest hospital.

APPLICATIONS

Public Telephone and Emergency Service

Even in a major city such as New York, there are many areas where it is impractical to provide conventional telephone service. Parks and other recreation areas, such as beaches, would require an extraordinary outlay of funds to be provided with the telephone coverage they should have. Ferries and passenger rail systems have no feasible solution for telephone service without cellular.

With cellular phone stations, public or emergency phones can be located anywhere within range of a cell site. These may be permanent installations or, in the case of special events like a Fourth of July celebration or a park concert, can be moved around to different areas as they are needed. They can also be invaluable for temporary telephone and emergency service in case of a natural disaster.

Often, emergency phone stations will not resemble telephones; rather, they will look just like emergency police or fire call boxes. Should an emergency arise, all a civic-minded citizen has to do is push a button to be connected with an assistance center.

Cellular phones are ideal for this type of application because each phone's numeric assignment module (NAM) contains identifying information unique to that phone. An assistance operator does not need to ask, "Where's the fire?" Using a computer to match a phone's electronic serial number (ESN) with its physical location, the operator can automatically pinpoint the emergency area. It is even possible to route the call directly to the closest fire, police, or other appropriate emergency service.

There are many parts of the country and the world where there is only sparse phone service or no service at all. Such areas include long stretches of some highways running through areas of low population density, and national and state parks.

Despite the obvious advantages of being able to summon emergency aid in such areas, it has previously been too impractical or expensive to provide conventional phones. Cellular phones change all this. Without the need for cables to a switching office, rugged phones

Use of cellular phone in an emergency situation. (Photo courtesy of Motorola, Inc.)

can be placed wherever they will be needed and left untended except for routine maintenance checks. In many parts of the country, you don't even need electricity, as the phones can be powered by self-contained battery packs, and recharged during the day by solar cells.

Underdeveloped countries as well as modern ones are finding that cellular offers a way to rapidly add telecommunications service where it is currently congested, slow to expand, or nonexistent. Cellular phones can be designed to connect to regular wireline phone extensions, provide a simulated dial tone, and operate for all intents and purposes like a wireline system. As a country develops economically, a mix of mobile and fixed operation can be accommodated with no change to the design of the system.

Wireless communications technology promises to displace many forms of fixed communications, as technology becomes more advanced and permits the flexibility of communications that an increasing population of users demand, at reduced costs approaching those of traditional wire communications.

GLOSSARY

A Carrier   The nonwireline carrier. See *Nonwireline*.

A-B Switch   A switch on a cellular phone that permits you to change between the A and B block carrier, or make one a priority. Usually the carrier you subscribe to will set their block as priority on your phone. Usually only your dealer or carrier will need to change this setting.

Access Charge   The fixed monthly charge for cellular subscription. This may include an allowance of airtime usage, as well as the use of special features like voice messaging, in some carriers' rate plans.

AMPS   An acronym used as the name for the current cellular standard used in the United States. It originally stood for Advanced Mobile Phone Service, the name AT&T chose for cellular service. The previous standard for mobile phone service, for example, had been called IMTS, or Improved Mobile Telephone Service. Digital cellular standards, like CDMA and TDMA, are also part of the AMPS standard.

Antenna   A length of wire that radiates or captures radio signals.

Area Code   The first three digits of your phone number (formerly the middle one of which was always 1 or 0, but now may be any digit) that identify the geographic area of telephone service. The local calling area of your cellular service may include more than one area code. The area code is technically referred to as an NPA, or Number Plan Area.

Band A portion of the radio frequency spectrum. Cellular communications take place in the 800-MHz region of the UHF band, while PCS is in the 1900-MHz area.

Base Station A cell site.

B Carrier The wireline carrier. See *Wireline.*

Block A group of radio frequencies within a band set aside for a particular purpose. Cellular telephony uses four blocks of frequencies in the 800-MHz portion of the UHF band. A-block and B-block carriers are assigned separate blocks of frequencies.

Burn (a NAM) To program information into the NAM (Numeric Assignment Module) of a cellular phone. The device that does this is a NAM burner. This process is being replaced by the ability to program the NAM directly through the phone.

Carrier The company that builds and operates the cellular system in a particular metropolitan or rural area. There are two cellular carriers in each area, and up to six PCS carriers may operate in the same area.

CDMA Acronym for Code Division Multiple Access. A digital cellular technology and standard that assigns a unique code to digitized segments of calls, allowing them to use vacant channels in a broad set of frequencies. This allows great economy in the use of channels, or radio-frequency spectrum. The digitized segments are reassembled using the code at the other end and converted back into analog voice for the listener. Compare with *TDMA*, a competing digital technology and standard.

CDPD Acronym for Cellular Digital Packet Data. A technology for digital data transmission using idle analog cellular channels. It permits cellular carriers to offer wireless point-to-point packet data services at speeds up to 9600 baud.

Cell The area covered by the transmitter/receiver of a cell site. A cell site may sectorize its antennas to service several cells from one location.

Cellphone Shortened form of cellular telephone. The term originated in Great Britain.

Cell Site The facility housing the transmit/receive radios and antennas and other equipment for one cell of a cellular system. The only real radio link in a cellular system is between the cellular telephone and the cell site; the rest of the network uses landline telecommunications facilities, although some of these may be microwave radio.

Cell Splitting   The process of converting one cell to multiple smaller cells by sectorizing antennas of a cell site or constructing additional cell sites within a cell. The power of each new cell is reduced to avoid interference with other cells using the same channels, and additional channel capacity and better coverage are obtained within the cell.

Cellular   Mobile radio telephony technology that uses multiple, small radio transceivers instead of one larger one to cover a geographical area. This allows frequency or channel reuse nearby within an area without interference, permitting greater system capacity and economical use of the available radio-frequency spectrum. The system requires a central computer to switch calls in progress from one cell to another as the user moves. The term is derived from the conceptual honeycomb-like pattern of coverage areas within a service area.

Cellular Telephone   The subscriber unit for cellular voice telecommunications.

Channel   A band of radio frequency wide enough (30 kHz) to carry a cellular conversation. Usually refers to a pair of frequencies, one for the cellphone-to-cell site (mobile-to-land) link, and one for the cell site-to-cellphone (land-to-mobile) link.

CLEAR (CLR)   The key on a cellular phone pressed to erase information from the display.

Connect Time   The period your cellular phone is in radio contact with a cell site, not to be confused with the actual length of time your conversation lasts. Connect time is measured from the time your phone's IN USE indicator lights up until you press the END key and it goes off, also called SEND-to-END time.

Control Head   The part of a mobile telephone installation, generally located near the handset, that acts as the "go-between" between the phone user and the transceiver/logic unit. In some phones, the control head *is* the handset.

Control Signal   A signal sent by a cell site to a cellular phone, or vice versa, carrying information necessary to the operation of the two, but not including the audio portion of a conversation. The channels used for control signals, called control channels, are separate from those used for voice. Control signals also flow between a mobile phone's handset and its transmitter/logic unit.

CPE Acronym for Customer Premises Equipment. A term that originally referred to landline telephones, but now refers to any subscriber telephone equipment, whether cellular or landline, on the premises or not.

DCS-1800 Acronym for Digital Cellular System. A GSM-based standard for PCS operating in the 1800-MHz frequency range outside the United States.

DCS-1900 Acronym for Digital Cellular System. A GSM-based standard for PCS, the same as DCS-1800, except operating in a higher frequency range (1900 MHz) for use in the United States.

DTMF Acronym for Dual Tone Multiple Frequency, better known by most people as the tones used in Touch Tone service. The tones generated when the keys are pressed on a telephone, used for telecommunications signaling, voice response systems, and voice messaging, as well as dialing.

Dual NAM A feature of some cellular telephones that permits them to be registered on more than one cellular system as the home system.

Dual Mode A feature of a cellular phone that permits both digital and analog operation.

END The cellular telephone key that terminates a call.

ESN Acronym for Electronic Serial Number. The serial number of a cellular phone, programmed into the telephone's NAM and used in combination with the subscriber telephone number by the cellular system to identify a user.

GSM Acronym for Groupe Speciale Mobile or Global System for Mobile communications. A standard for digital cellular originally created to unite the various uncoordinated systems of individual countries under a single standard in Europe, but which has become the most widely accepted standard for cellular everywhere in the world except the United States and Japan. Groupe Speciale Mobile was the term for the original standards group. The second term was created artificially to give the acronym some public relations meaning.

Handoff The transfer of control of a call in progress from one cell site to another, changing the channel assignment as well.

Handset The portion of a transportable or mobile cellular telephone that the user holds, including the microphone and miniature speaker, and usually incorporating the dial pad. The handset is integrated with the rest of the phone in a portable.

Hands-Free Operation of the cellular phone without using the handset, usually referring to use in a vehicle. A special microphone and speaker are installed for this purpose. Portables can be adapted to hands-free mode by using a special cradle for insertion of the telephone while in the vehicle.

HORN A function on mobile-installed telephones that when activated beeps the horn to signal an incoming call, and allows the phone to stay on with the ignition off. Also called an *Auxiliary Alert* or *Call-in-progress Protection*.

IN USE An indicator on cellular phones that tells the user that a call is in progress.

Keypad A set of push-button electronic switches. The keys on a calculator make up a keypad, as do the buttons on a cellular phone.

kHz Abbreviation for kilohertz. A measure of radio frequency. One kHz is one thousand cycles per second.

Landline Traditional, wire-based telephony, used to distinguish it from mobile telephony.

LOCK A function on cellular telephones that, when activated, prevents use of the telephone until the user enters a security code.

MHz Abbreviation of megahertz. A measure of radio frequency. One MHz is one million cycles per second.

Modem Acronym for MOdulator/DEModulator. A device used to send information from one computer to another over a voice telephone line.

MSC Acronym for Mobile Switching Center. See *MTSO*.

MTSO Acronym for Mobile Telephone Switching Office. The switch that controls call setup, channel allocation, user features, and cell-site assignment of cellular telephony, and links the cellular system to the landline telephone system. Now usually referred to as an MSC (Mobile Switching Center).

NAM Acronym for Numeric Assignment Module. A chip in a cellular telephone that contains its ESN and telephone number assignment; the latter is programmed at the time the user subscribes.

Nonwireline Term referring to frequency block A cellular carriers, because block B carriers were originally all conventional telephone companies. Now simply referred to as the A carrier.

NO SERVICE An indicator on a cellular phone that tells users that they are in an area where cellular service is unavailable, either because the signal is weak, or they are outside the coverage area of a system.

Off-Peak A calling period defined in cellular rate plans, referring to times of day when airtime rates are less because the cellular system is not as busy as during peak calling times. Usually refers to 7 P.M.–7 A.M. weekdays as well as weekends and holidays.

PCS Acronym for Personal Communications Services. A family of cellular-like voice telecommunications services (as well as nonvoice services like paging) on a different frequency and power level than cellular. There is no single standard defining the capabilities and features for PCS in the United States, like AMPS for cellular, and the term only defines the frequencies and licenses set aside by the FCC.

Peak A calling period defined in cellular rate plans, referring to times of day when airtime rates are higher because the cellular system must be built for higher call volume. Usually refers to 7 A.M.–7 P.M. weekdays.

PIN Acronym for Personal Identification Number. A code used when making a cellular call to deter fraudulent use.

RCL A cellular telephone function that recalls a telephone number from memory.

Reuse The assignment of frequencies or channels to cells so that adjoining cells do not use the same frequencies and cause interference, yet cells out of range of one cell can use the same ones; this expands the capacity of the system by enabling it to use the same channel in many areas, or cells, throughout the system simultaneously.

ROAM An indicator on the cellular phone that a carrier other than the home carrier is providing service. When flashing, it means that the alternate carrier is on the opposite frequency block of the home carrier.

Roaming Using a cellular telephone outside the service area of the home carrier.

SEND (SND) The cellular telephone key that initiates a call, or answers an incoming call.

SMS Acronym for Short-Message Service. A wireless network feature that permits short messages to appear on the display of specially equipped cellular and PCS phones.

Subscriber One who receives cellular service from a carrier in return for a monthly fee under a service agreement; may be distinguished from the actual user of a phone on a particular call.

STO A cellular telephone function that stores a telephone number in memory.

TDMA Acronym for Time Division Multiple Access. A digital cellular technology and standard that digitizes portions of the call and assigns them to specific time slots on a single channel, permitting several conversations on one channel. Compare with *CDMA*, a competing digital technology and standard.

Transceiver A combination radio transmitter/radio receiver. A cellular telephone or cell site radio is a transceiver.

Voice Messaging The recording, storing, and retrieval of voice telecommunications transmissions. It is used in cellular service primarily as a telephone answering capability, especially since cellular users often do not have their telephone turned on or are out of the coverage area.

Wireless The broad term for radio telecommunications that includes cellular, PCS, paging, wireless data, and other mobile voice and data services.

Wireline Term referring to frequency block B cellular carriers. Originally they were all conventional telephone companies. Now simply called the B carrier.

INDEX